T0075491

# THE SOLAR SYSTEM

# THE SOLAR SYSTEM

## EXPLORING THE SUN, PLANETS AND THEIR MOONS

ROBERT HARVEY

First published in 2022



Published by Amber Books Ltd
United House
London N7 9DP
United Kingdom
www.amberbooks.co.uk
Facebook: amberbooks
Instagram: amberbooksltd
Twitter: @amberbooks
Pinterest: amberbooksltd

ISBN: 978-1-83886-169-8

The author is very grateful to Gavin James and Kate Eiloart for reviewing the text and making many helpful suggestions.

Editor: Michael Spilling
Designer: Mark Batley
Picture research: Terry Forshaw

Printed in China

# CONTENTS

# Introduction

THE LAST 60 YEARS have been an unprecedented period of discovery. On 14 December 1962, the space probe *Mariner 2* flew past Venus, the first human-made object to reach another planet. This achievement heralded the start of the most ambitious and far-reaching period of exploration in human history.

The name 'Mariner' for that intrepid spacecraft was a reference to the great seafaring voyages of past centuries that explored distant lands. In the early eighteenth century, much of the Pacific Ocean was unknown to Europeans.

Maps of the world showed empty space or hypothetical land masses. Great voyages of the later eighteenth century filled in the gaps on maps of the world. The most renowned navigator, Captain James Cook, mapped islands and continents in three voyages to the Pacific between 1768 and 1779. Scientific studies conducted on these voyages revealed hitherto unknown wonders of our planet.

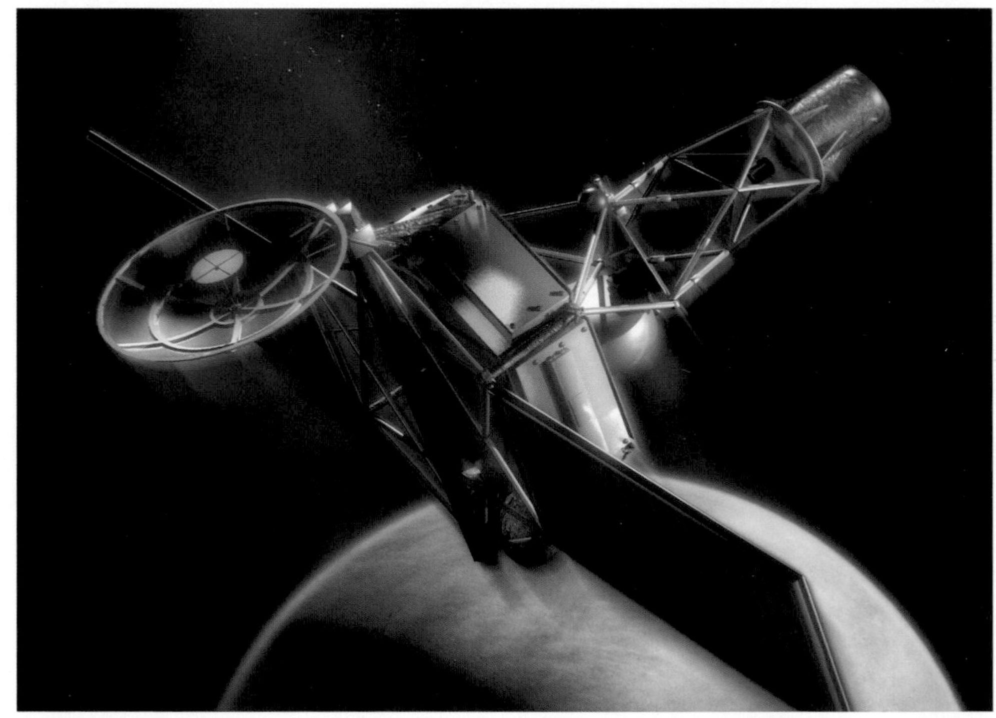

ABOVE:
**Artist's concept of the *Mariner 2* space probe passing Venus.**

OPPOSITE:
**Colour maps of Jupiter constructed from images taken by the narrow-angle camera onboard NASA's *Cassini* spacecraft.**

In much the same way, in the first half of the twentieth century our Solar System was little known. With only hazy views from Earth-bound telescopes as evidence, science-fiction writers were free to speculate about the existence of alien civilizations on Venus and Mars. The exploration of the Solar System heralded by *Mariner 2* has filled in our knowledge. This endeavour, the modern equivalent of eighteenth century seafaring voyages, has for the most part been undertaken not by human travellers but by robots.

Within the span of a human lifetime, our spacecraft have visited all eight planets of the Solar System, together with several dwarf planets, asteroids and comets. We have mapped the surface of Mercury and Venus in exquisite detail, landed rovers on Mars, placed orbiters around Jupiter and Saturn, and parachuted to the surface of Titan. Our emissaries have visited icy worlds five billion kilometres from home and continued onwards to reach interstellar space. The pictures and science returned by these intrepid travellers have transformed our understanding of the Solar System in which we live.

Amongst the most fascinating aspects of our exploration of the Solar System has been finding how and why other planets are both similar to and different from Earth. Many features and processes familiar on Earth occur elsewhere; yet our planet is radically different from all the others.

ABOVE:
***Ingenuity*** **on the surface of Mars. In April 2021, 118 years after the Wright brothers made the first powered flight on Earth, this small helicopter made the first powered flight on another planet.**

OPPOSITE:
**Launch of the *Parker Solar Probe* on 12 August 2018.**

Arguably it is not the planets themselves but their moons that have created the greatest surprises. Before the age of space flight, no one suspected that Europa has an underground ocean of liquid water, Titan has seas of liquid methane, Io has erupting volcanoes or Enceladus has active geysers. Dwarf planets have proved similarly intriguing; our visit to Pluto revealed mountains and tectonic plates made of solid ice.

This book is a story of fantastic exploration, using the images sent by our robot spacecraft to show the extraordinary diversity of our Solar System.

## The Planets

There are eight planets in the Solar System. They are defined as spherical bodies in orbit around the Sun, whose gravitational influence has cleared their orbit of other objects. The planets can be divided into two types: four terrestrial planets (which are rocky with a solid surface like Earth) and four giant planets (which comprise mostly gas, ices or volatiles and have no solid surface). Terrestrial planets are relatively small, close to the Sun and have few or no moons; whilst giant planets occupy the outer Solar System and have many natural satellites.
All eight planets are thought to have formed around the same time, 4.6 billion years ago, from a disc of dust and gas surrounding the young Sun.

| Planet | Average distance from the sun | | Orbital period (2) | Equitorial diameter (km) | Rotational period (3) | Number of known moons (4) |
|---|---|---|---|---|---|---|
| | Million km | Astronomical units (1) | | | | |
| Mercury | 58 | 0.39 | 88 days | 4879 | 59 days | 0 |
| Venus | 108 | 0.72 | 225 days | 12,104 | 243 days | 0 |
| Earth | 150 | 1 | 365 days | 12,756 | 23.9 hours | 1 |
| Mars | 228 | 1.5 | 687 days | 6792 | 24.6 hours | 2 |
| Jupiter | 779 | 5.2 | 11.9 years | 142,984 | 9.9 hours | 80 |
| Saturn | 1433 | 9.5 | 29.5 years | 120,536 | 10.7 hours | 83 |
| Uranus | 2872 | 19 | 84 years | 51,118 | 17.2 hours | 27 |
| Neptune | 4495 | 30 | 165 years | 49,528 | 16.1 hours | 14 |

**Notes:**
(1) One astronomical unit is the average distance from Earth to the Sun.
(2) Throughout this book, the time units 'days' and 'years' are those of Earth, unless the term is qualified to refer
to the day or year of another planet (e.g. 'a Martian year').
(3) Rotation period is expressed relative to the stars. Hence Earth rotates once in 24 hours
relative to the Sun but once in 23 hours, 56 minutes and 4 seconds relative to the stars.
(4) The capitalized term 'Moon' means Earth's Moon, whilst the uncapitalized term 'moon' refers to the natural satellites
of other planets.

## SIZE OF THE PLANETS IN RELATION TO EACH OTHER

SATURN
URANUS
NEPTUNE

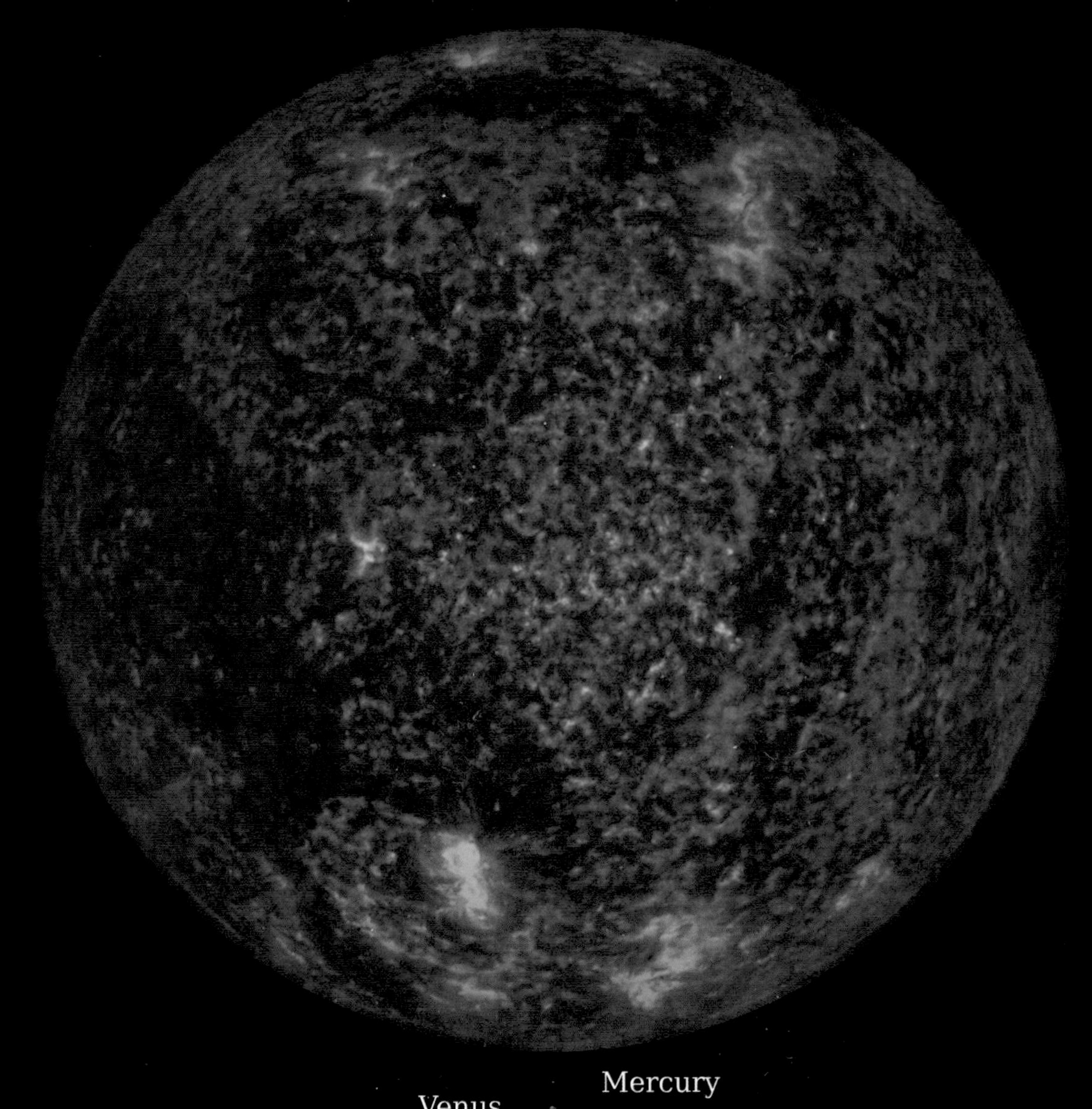

Mercury
Venus
Earth
Mars
Jupiter

Saturn

**Relative distances of the planets from the Sun.**

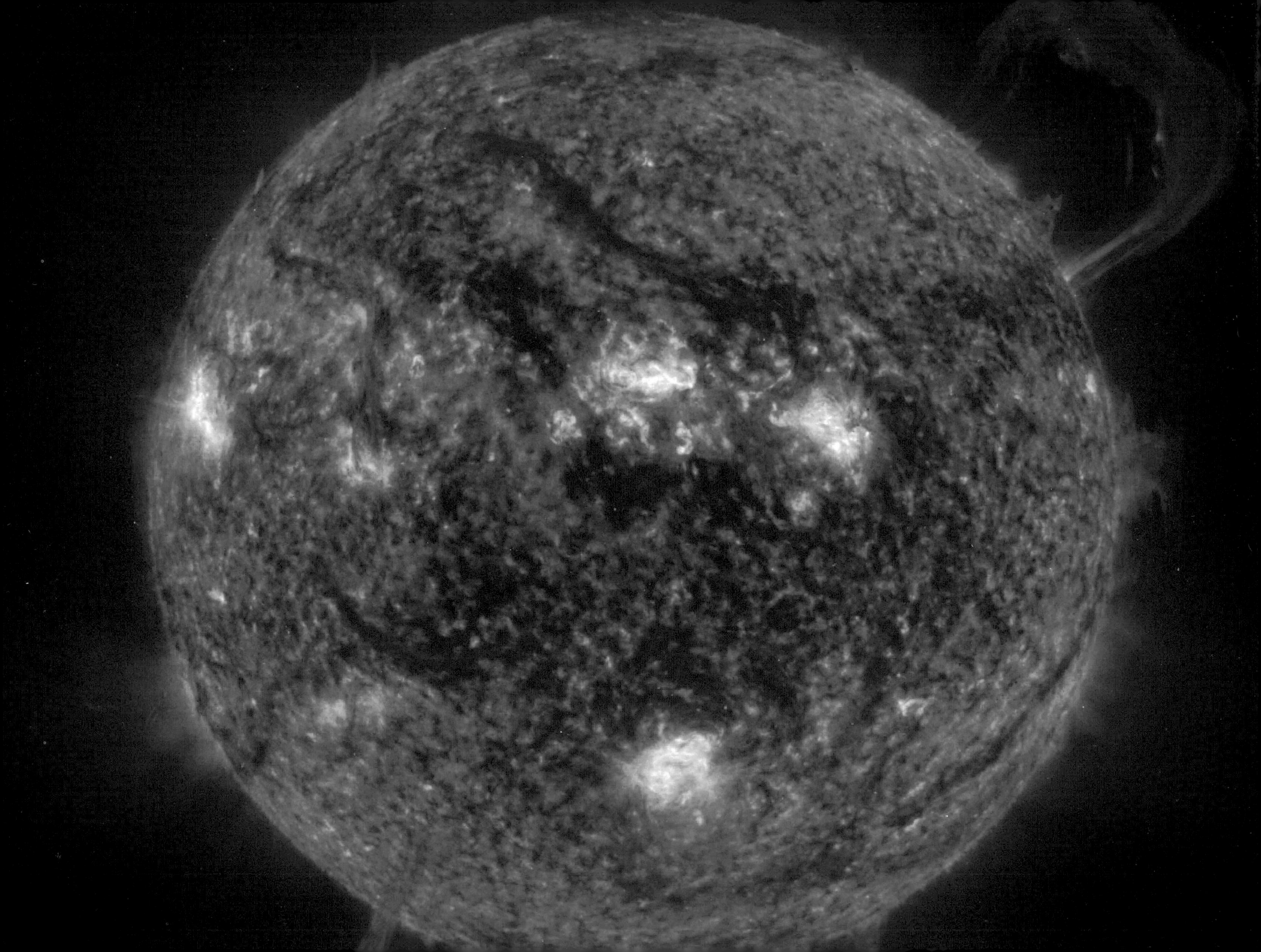

# The Sun

The Sun dominates our Solar System. Its huge gravitational pull keeps the planets and almost everything else in orbit around it. The Sun brings light and warmth to our part of space. It is the foundation of nearly every food chain on Earth.

However, the Sun's source of energy was discovered only a little more than a century ago, by physicist Arthur Eddington. Drawing on Albert Einstein's famous equation $E=mc^2$, he showed that the Sun is powered by nuclear fusion. Under conditions of extreme pressure and high temperature in its core, the Sun converts hydrogen into helium. Every second, 620 million tonnes of hydrogen are fused to make 616 million tonnes of helium. The remaining four million tonnes of mass is released as 385 septillion ($10^{24}$) watts of energy. With a temperature of some 15.7 million°C (28.2 million°F), the Sun's core represents about a quarter of its diameter. Energy generated in the core takes thousands of years to reach the surface, where it is emitted as visible light and other electromagnetic radiation.

Although the Sun is central to our Solar System and immensely important to Earth, in fact it is a fairly average star. One of about 250 billion stars in our galaxy, it is around halfway through its lifespan. As its fuel is consumed, the Sun is becoming brighter by about one per cent every 100 million years. Eventually, Earth will heat up so much that it loses its water.

Some five billion years from now, the Sun will run short of hydrogen in its core and evolve into a red giant. Its diameter will expand to engulf the orbits of Mercury, Venus and possibly Earth. Even if Earth escapes incineration, it will certainly be uninhabitable.

OPPOSITE & OVERLEAF LEFT:
**Solar prominences**
Prominences are huge, bright plumes of gas that extend many thousands of kilometres from the photosphere. When viewed against the surface of the Sun, they are referred to as filaments. They form in filament channels where magnetic polarity is reversed and may last for weeks or months. The longest recorded prominence (February 2022) extended three million km (two million miles) from the surface, which is twice the Sun's diameter. Prominences were first observed and photographed during total solar eclipses. They are also visible on H-alpha views of the Sun such as those shown here.

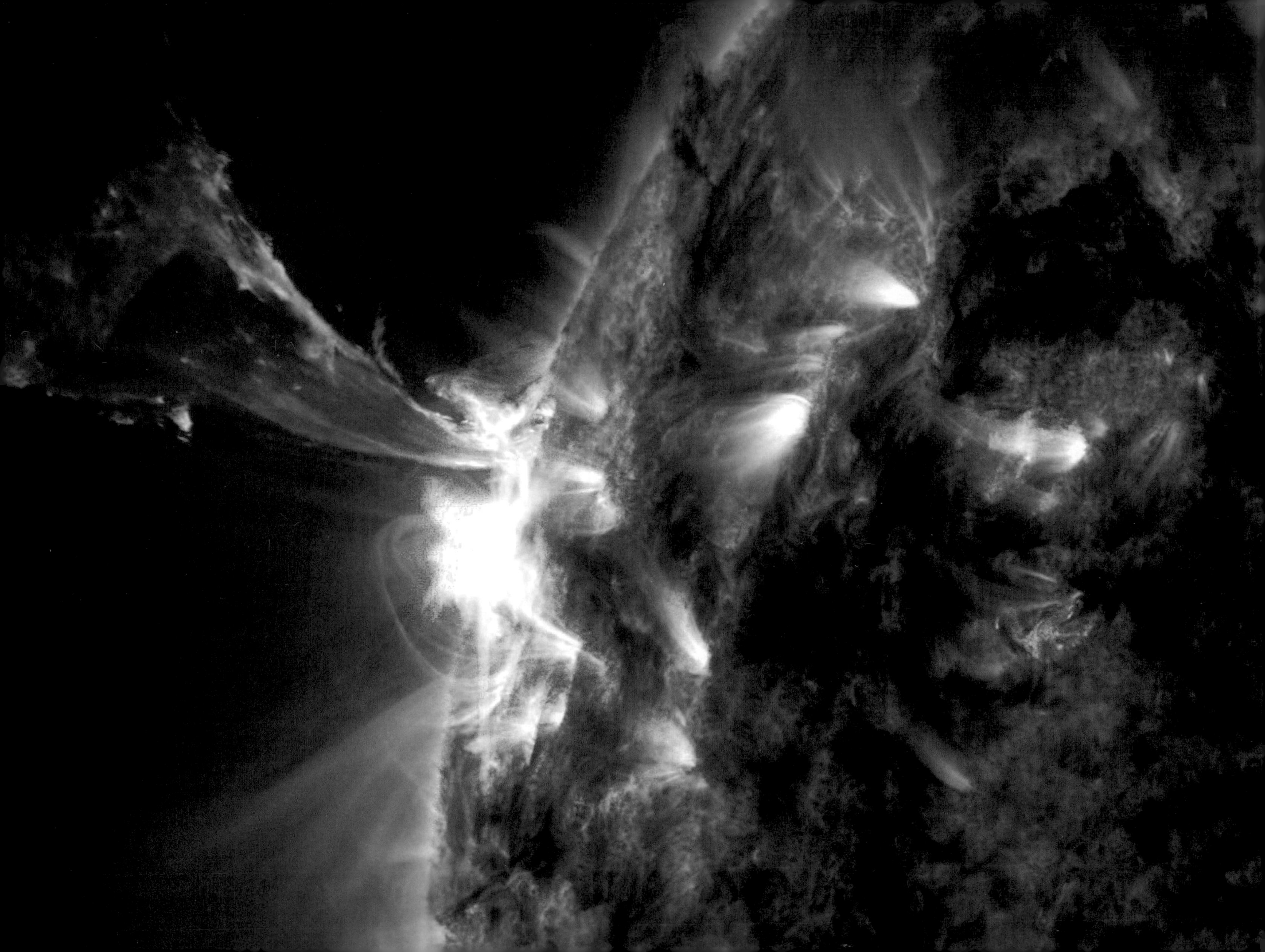

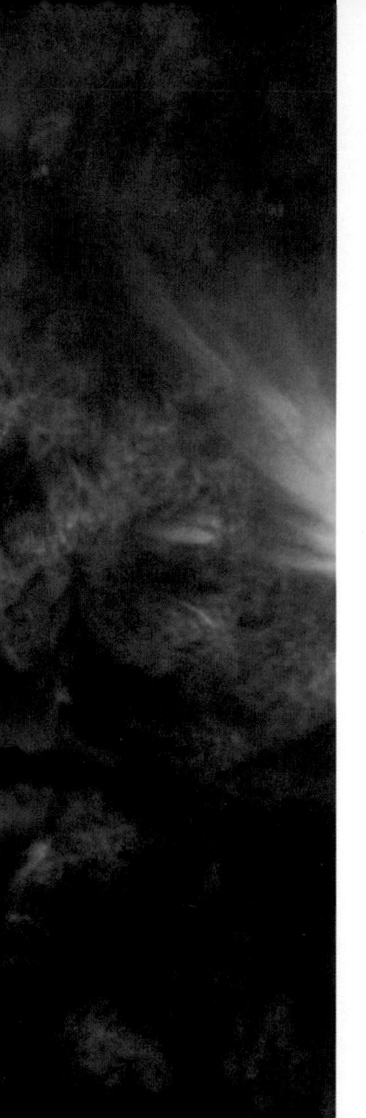

RIGHT:

**The Sun in hydrogen alpha**

Hydrogen, comprising one proton and one electron, is by far the most abundant element in the Sun. When it absorbs energy, the electron is temporarily raised to a higher orbit. This excited state is unstable so the electron drops back to its normal orbit, releasing its energy as light of a specific wavelength, 656 nanometres, which is towards the red end of the visible spectrum. The process of energizing hydrogen electrons occurs most efficiently within an outer layer of the Sun, referred to as the chromosphere.

A hydrogen alpha (H-alpha) filter isolates light of this characteristic wavelength and enables us to see detail in the solar chromosphere. The Sun no longer appears as a bright, plain disc. Instead we see a swirling surface of seething activity.

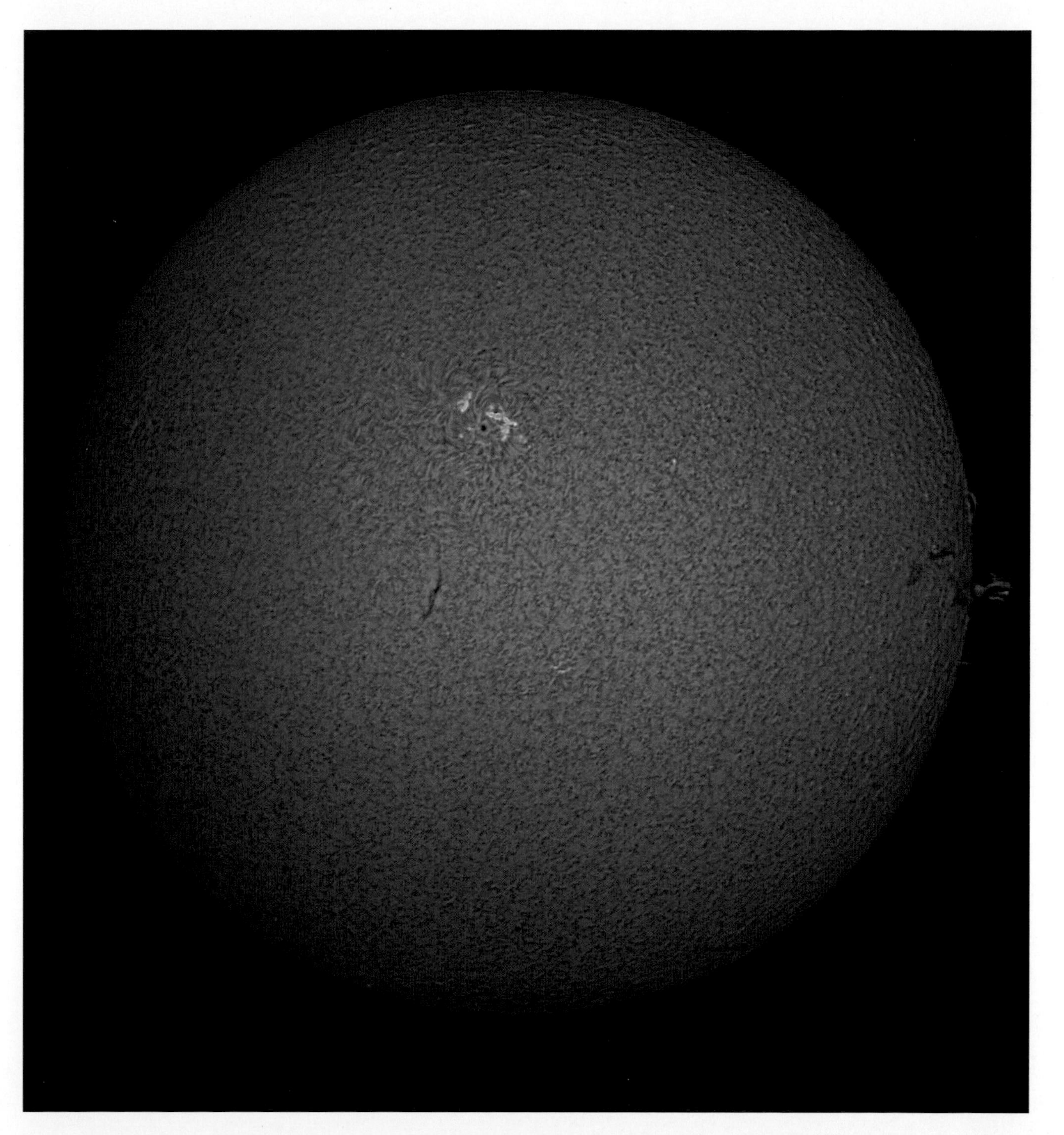

ALL:
**Solar transit of Venus**
It is not often that we get the opportunity to directly compare the size of the Sun to that of the planets in a single field of view. Venus is similar in size to Earth and very occasionally it can be seen transiting the face of the Sun, enabling the huge size of the Sun to be directly appreciated. The Sun is 1.39 million km (860,000 miles) in diameter, which is 109 times the diameter of Earth. In terms of volume, 1.3 million Earths would fit inside the Sun. Another measure of its dominance within our Solar System is that the Sun contains 99.86 per cent of the mass. The planets and everything else in this book make up just 0.14 per cent of the Solar System.

Transits of Venus occur in pairs eight years apart, separated from the next pair by more than a century. The last transits occurred in 2004 and 2012; the next two will be in 2117 and 2125. These very rare events were of great historical importance to science as they provided the first way for astronomers to measure the distance to the Sun. This in turn revealed the scale of the whole Solar System, which had previously been unknown. In 1761 and 1769 scientists travelled all over the world (including Captain James Cook to Tahiti) to observe two transits of Venus. By comparing their observations of the exact moment the transits began and ended as seen from widely spaced locations, the distance to the Sun was calculated to be 151.7 million km. This was within two per cent of the currently accepted value of 149.6 million km (93 million miles).

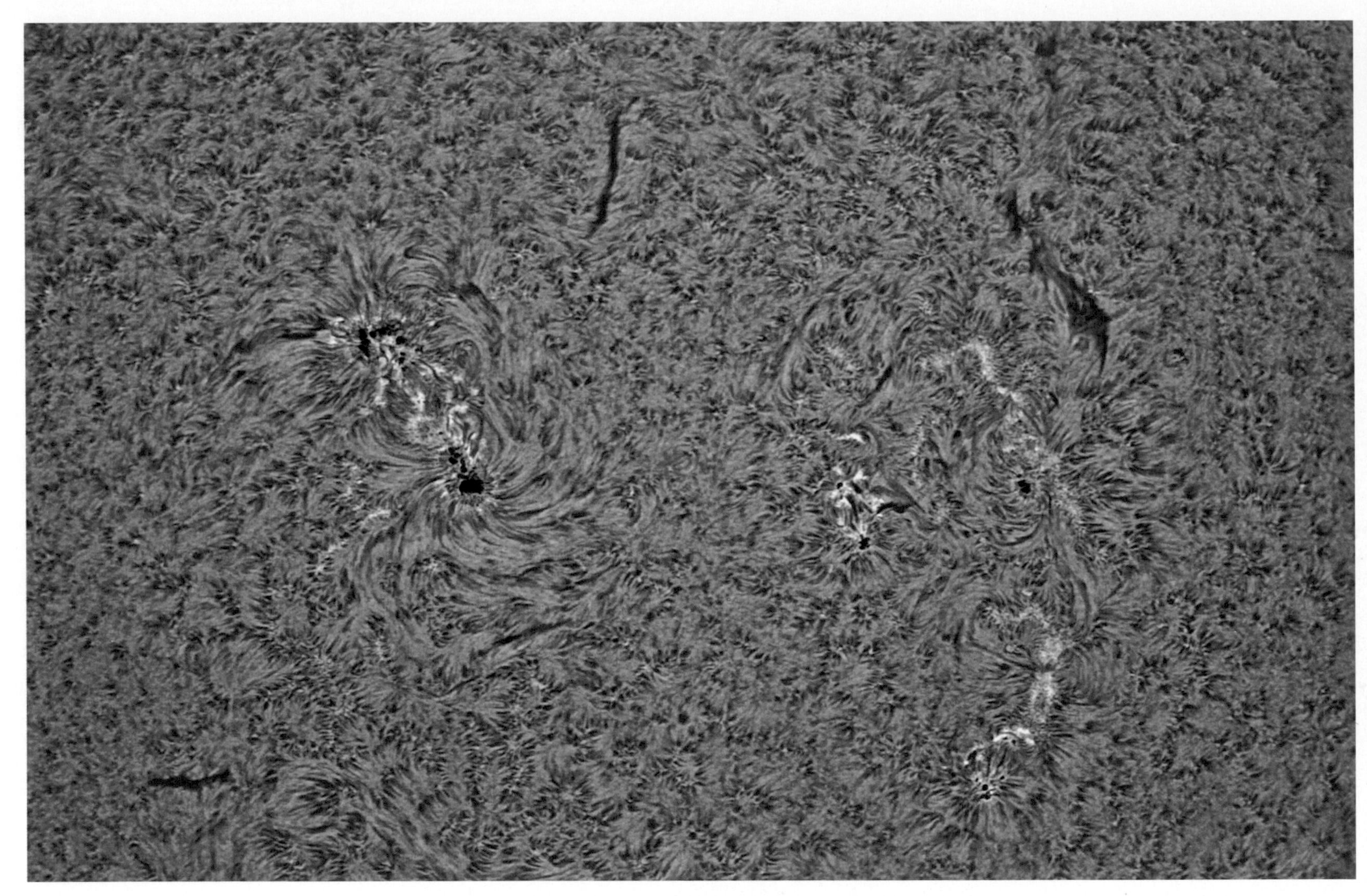

ABOVE:

**Detail of the Sun's surface**

The solar chromosphere, around 2000km (1245 miles) thick, is heated by nuclear fusion occurring in the Sun's core to temperatures exceeding 10,000°C (18,000°F). This intense heating causes convection, giving the chromosphere a mottled appearance in H-alpha as plasma rises and falls. Dark reddish spicules rise to the top of the chromosphere and then sink back down again over timescales of 10 to 15 minutes. Bright plages are active regions where magnetic flux breach the solar surface. These may develop into groups of sunspots. Bundles of fibrils are hot plasma confined in magnetic tubes. This picture was taken in December 2015, 20 months after the peak activity of an 11-year solar cycle.

RIGHT:

**X-Ray and ultraviolet view of the Sun**

Whilst H-alpha gives us a view of the Sun in a narrow band of visible light, in fact the Sun emits electromagnetic radiation at a huge range of wavelengths. Most are invisible to our eyes and some are dangerous to life.

This image is a mosaic incorporating observations in April 2015 from three different telescopes in Earth orbit, all of which photograph short wavelength radiation. High-energy X-rays recorded by *NuSTAR* are shown in blue, low-energy X-rays imaged by *Hinode* are in green and extreme ultraviolet light seen by the *Solar Dynamics Observatory* is yellow and red. The brightest areas are flaring, active regions.

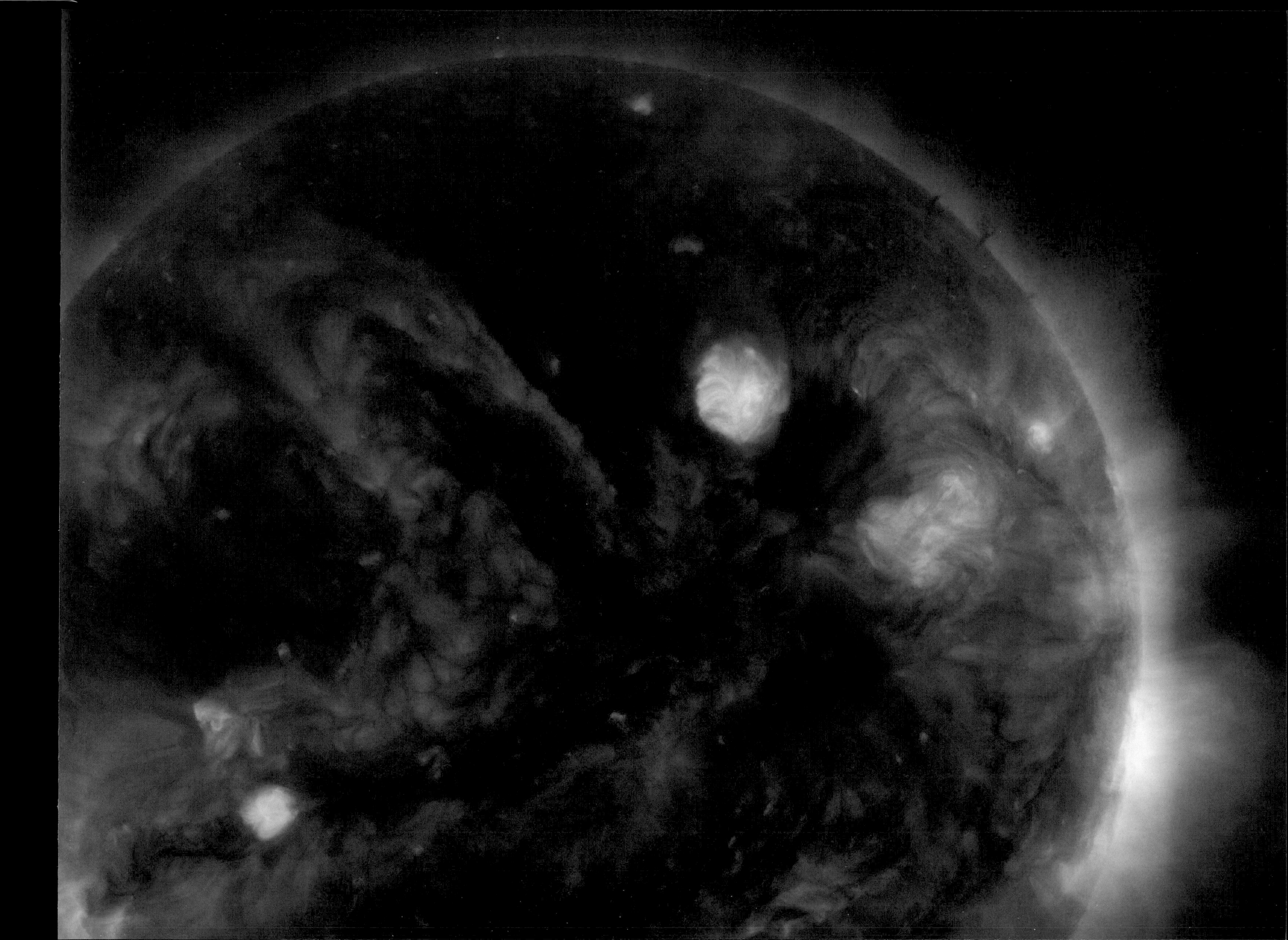

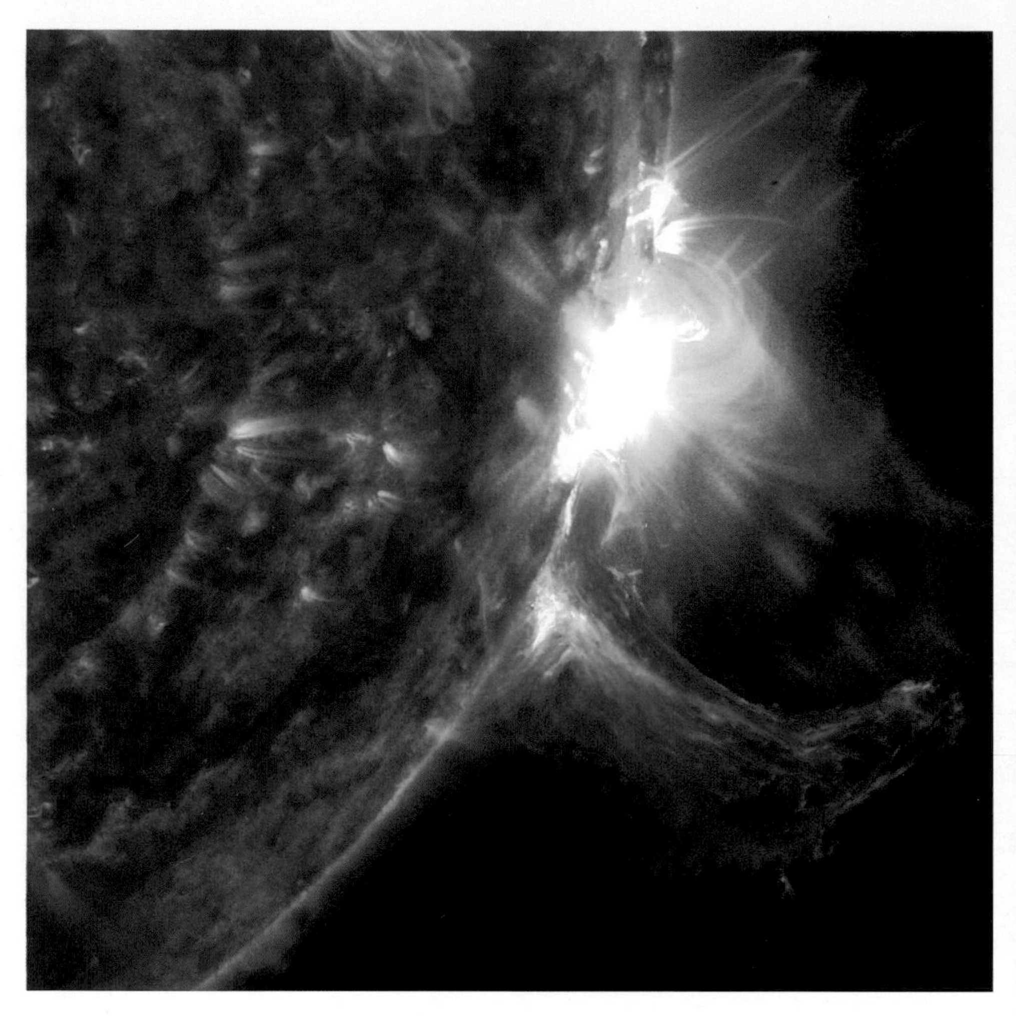

ABOVE:

## Solar flare

Solar flares are intense eruptions of electromagnetic radiation within the Sun's photosphere, chromosphere and corona. They are visible in H-alpha images of the Sun but most of their energy is emitted in other wavelengths ranging from radio waves to gamma rays.

Flares occur when charged particles such as electrons interact with plasma. Their frequency is closely linked to the Sun's 11-year magnetic cycle. At solar maximum several flares can be detected each day, whereas during solar minimum there may be fewer than one per week. The charged particles associated with a flare join the solar wind and may present radiation hazards to spacecraft outside Earth's protective magnetosphere.

RIGHT:

## Flux ropes on the Sun

Magnetic loops extending from the surface of the Sun are known as flux ropes. They are closely associated with the origin of coronal mass ejections. Large-scale flux ropes give rise to the strongest disturbances in Earth's magnetosphere.

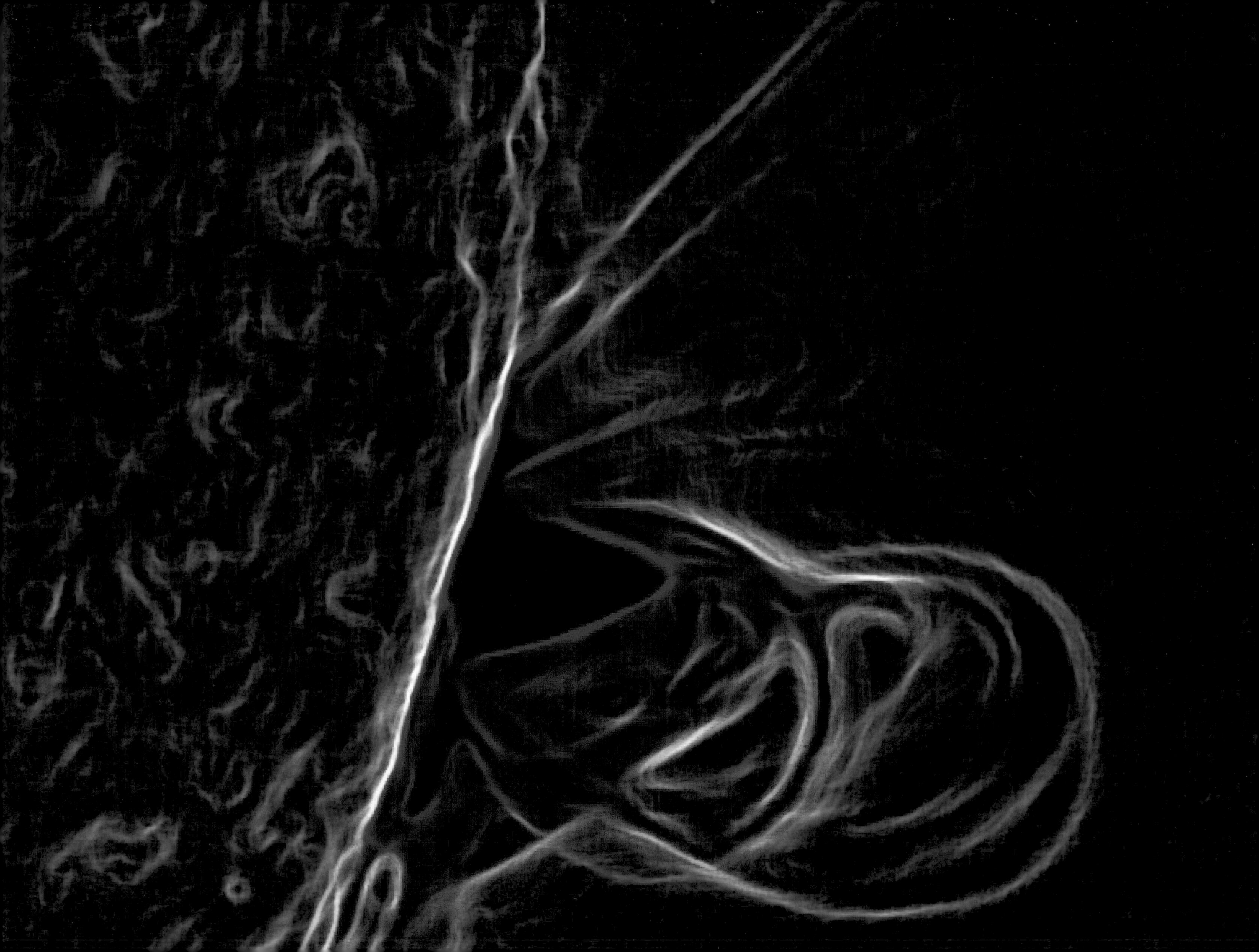

**Visiting the Sun**

Many observations of the Sun have been made from Earth-based telescopes. In 2018 a spacecraft was launched specifically to study our local star and understand the origins of the solar wind. *Parker Solar Probe* will approach closer than seven million kilometres from the Sun's centre, within the solar corona. To achieve this the probe makes seven close fly-bys of Venus, using repeated gravity assists to shrink its elliptical orbit. On 28 April 2021 it first entered the solar corona 13 million km from the Sun. The probe has measured frequent, short-term changes in the Sun's intense magnetic field which help us understand how the corona is heated.

When it makes its closest approach to the Sun (perihelion) in 2025, the probe will be travelling at a speed of up to 200km/sec (124miles/sec), making it the fastest human-made object. At this distance, Parker will experience intense solar radiation, around 475 times as strong as we do on Earth. A hexagonal solar shield 11.4cm (4in) thick with an aluminium oxide surface protects the scientific instruments from solar glare and outside temperatures of 1370°C (2500°F). This picture is an artist's impression of Parker Solar Probe approaching perihelion.

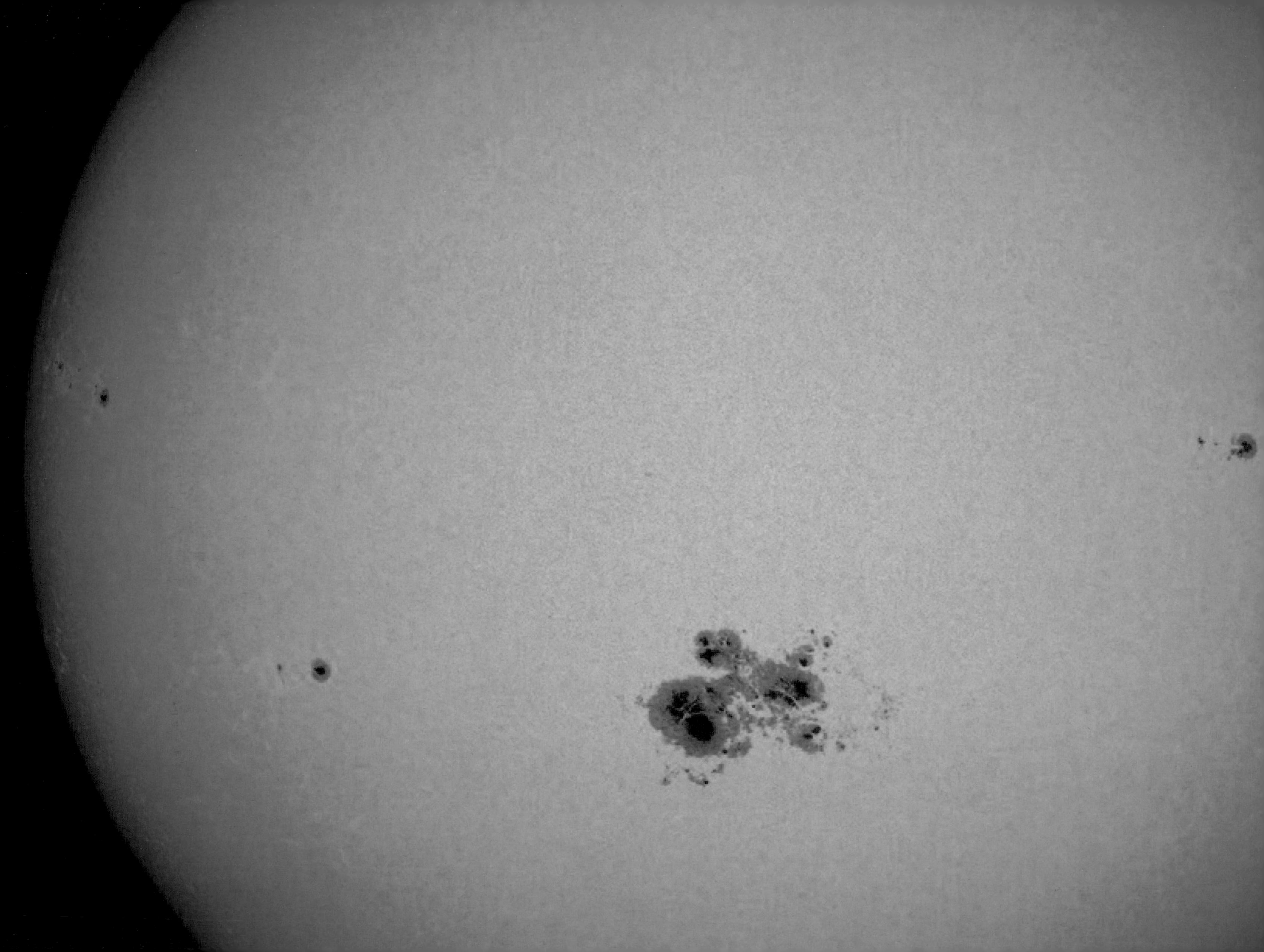

LEFT:

**Sunspots**

The photosphere is the visible surface of the Sun from which light is emitted. About 100km (62 miles) thick below the chromosphere, its temperature is around 5500°C (9900°F). Sunspots are areas of relatively cool photosphere, with a core temperature between 2700 and 4200°C (4900–7600°F). Relatively short-lived features, their typical lifespan is days, weeks or occasionally months. They are caused by concentrations of magnetic flux that inhibit convectional heating within the photosphere. Sunspots are linked to several other phenomena including flares, prominences and coronal mass ejections. As seen from Earth, they are carried across the face of the Sun as it rotates, which takes between 25 and 29 days to complete.

Observations since the seventeenth century have shown that the numbers and intensity of sunspots vary over a cycle of around 11 years. We now know that this reflects a fundamental solar cycle of magnetic activity. The Sun reverses its magnetic polarity as sunspots reach their maximum. The last solar cycle peaked in 2014 and the current one is expected to reach another maximum between 2023 and 2026. The solar cycle affects Earth in a number of ways, including the arrival of charged particles from solar storms (which can affect global communications systems), the occurrence of aurorae and even our climate.

Pictured left is the largest sunspot observed during the last solar cycle, with a surface area about 16 times that of the Earth.

RIGHT:

**Coronal mass ejection**

A coronal mass ejection (CME) is a release of plasma and associated magnetic field from the solar corona. They result from magnetic field lines becoming twisted, leading to a build-up of energy. A CME is a release valve for that energy. Coronal mass ejections are associated with sunspots, flares and may also arise from a solar prominence that breaks off and is released into space.

The ejected matter in a CME comprises highly energetic electrons and protons. When these are directed towards Earth they causes a geomagnetic storm. Spectacular aurorae may be seen in Earth's upper atmosphere, not only around the Arctic and Antarctic but also in mid-latitudes. In extreme cases, a geomagnetic storm may lead to radio blackouts and power grid failures, such as occurred on 13 March 1989. The biggest geomagnetic storm ever recorded was in September 1859, resulting in spectacularly bright aurorae worldwide and the interruption of telegraph communication systems. Modern technology makes human society much more vulnerable to a storm of this magnitude if it occurred today.

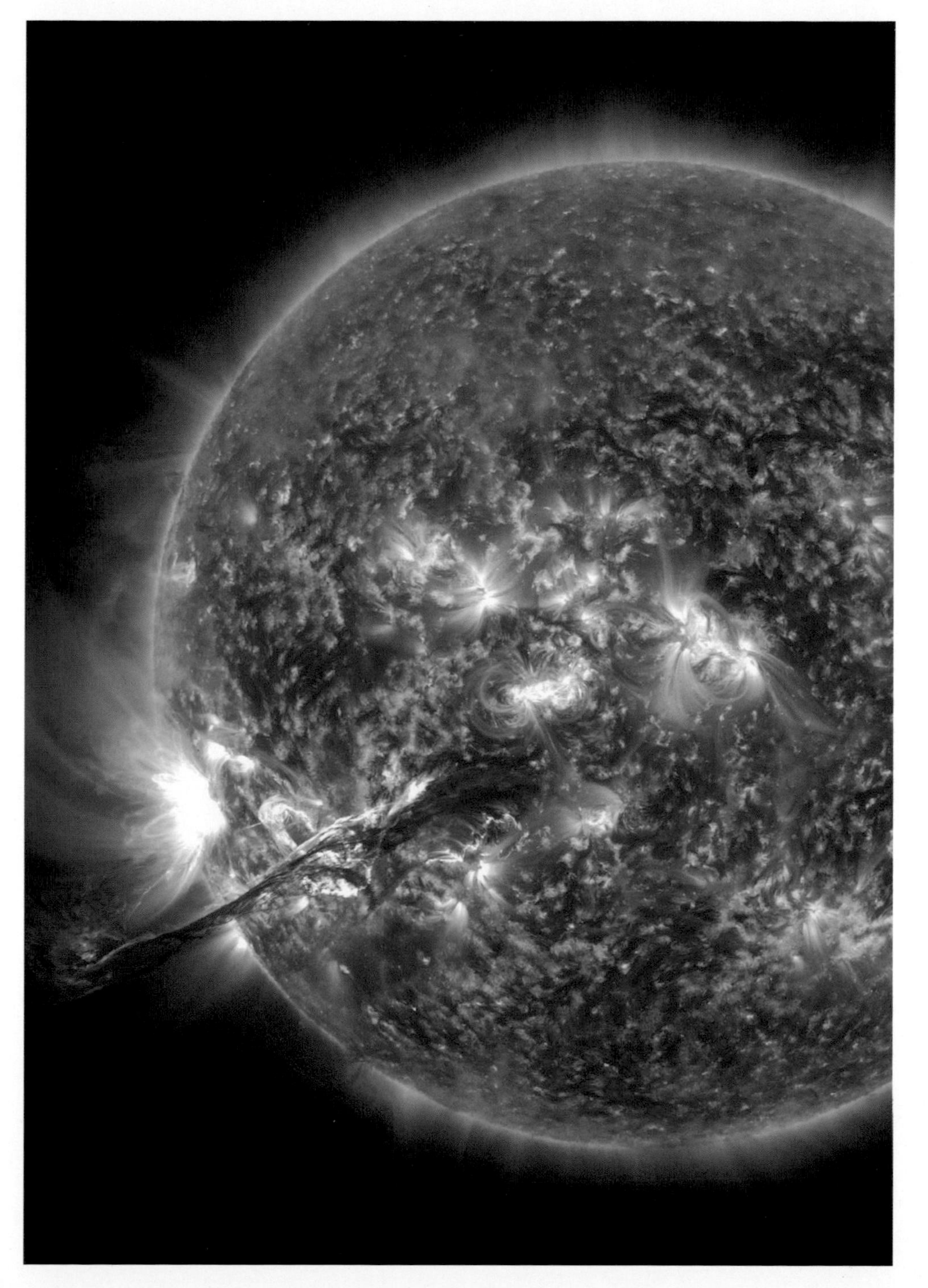

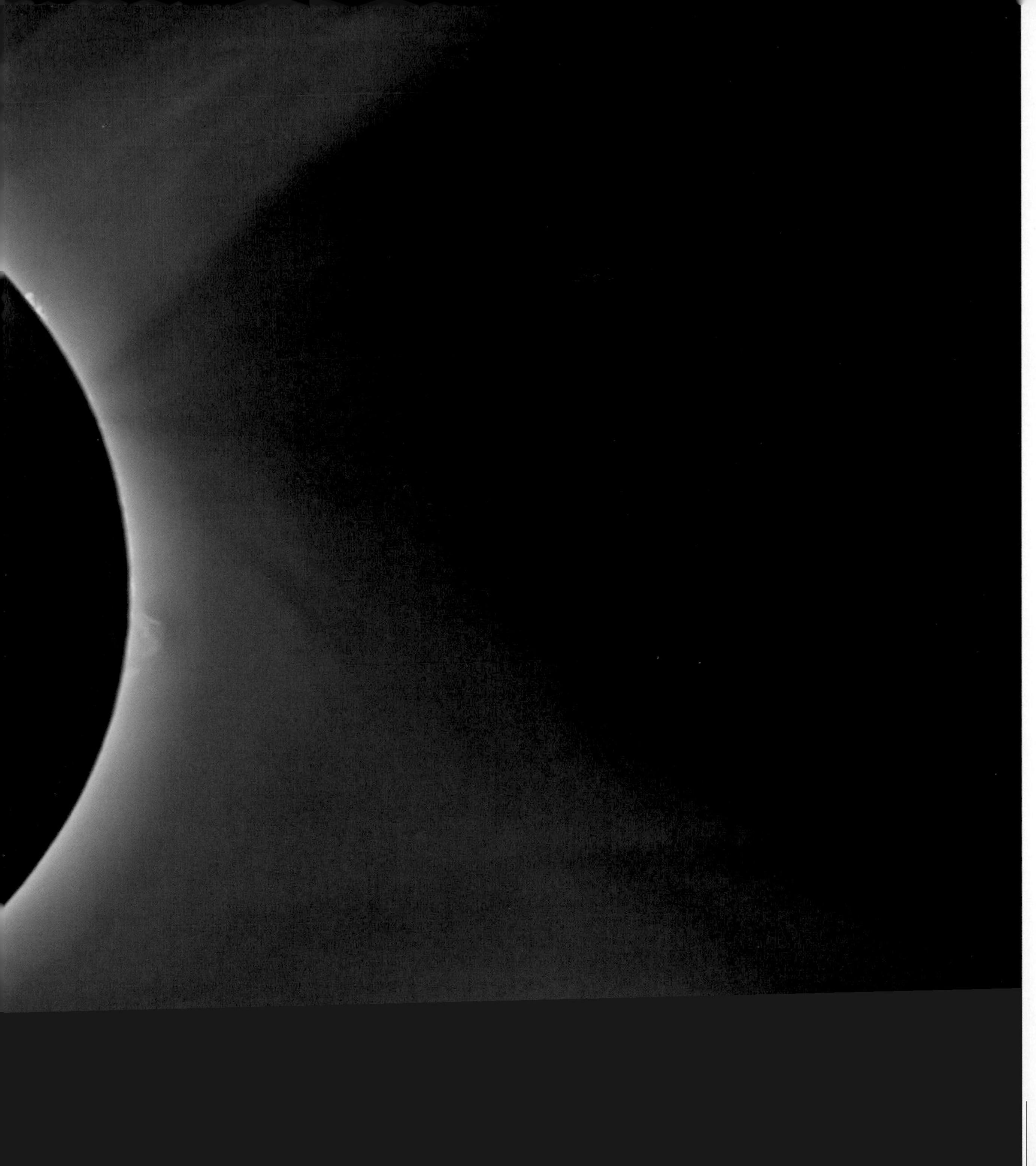

**Solar corona**

The Sun's outer atmosphere is called the corona, which extends millions of kilometres into space. It is a trillion times less dense than the photosphere and about one million times fainter. Hence, our views of the corona are obscured by the brightness of the photosphere. During a total solar eclipse the photosphere is hidden by the Moon, revealing the glory of the corona.

The corona is a plasma with a temperature of around one million degrees Centigrade, which is hundreds of times hotter than the photosphere. The mechanism of this heating is not fully understood. Because atoms in plasma are ionised (stripped of an electron), they are electrically conductive. Structure in the corona is the result of its ions being controlled by the Sun's intense magnetic field. In this photograph (left) taken during the total solar eclipse of 21 August 2017, the position of the Sun's north and south magnetic poles at the top and bottom can clearly be discerned. During the peak of the solar cycle, the corona exhibits complex large-scale structure.

Charged particles are constantly emitted from the corona into space. This solar wind comprises protons, electrons and alpha particles (helium nuclei) and extends throughout the Solar System. The effective outer limit of the solar wind is called the heliopause. The spacecraft *Voyager 1* passed through the heliopause at a distance of around 18 billion kilometres (11 billion miles) from the Sun, so we know the solar wind reaches 150 times further than Earth.

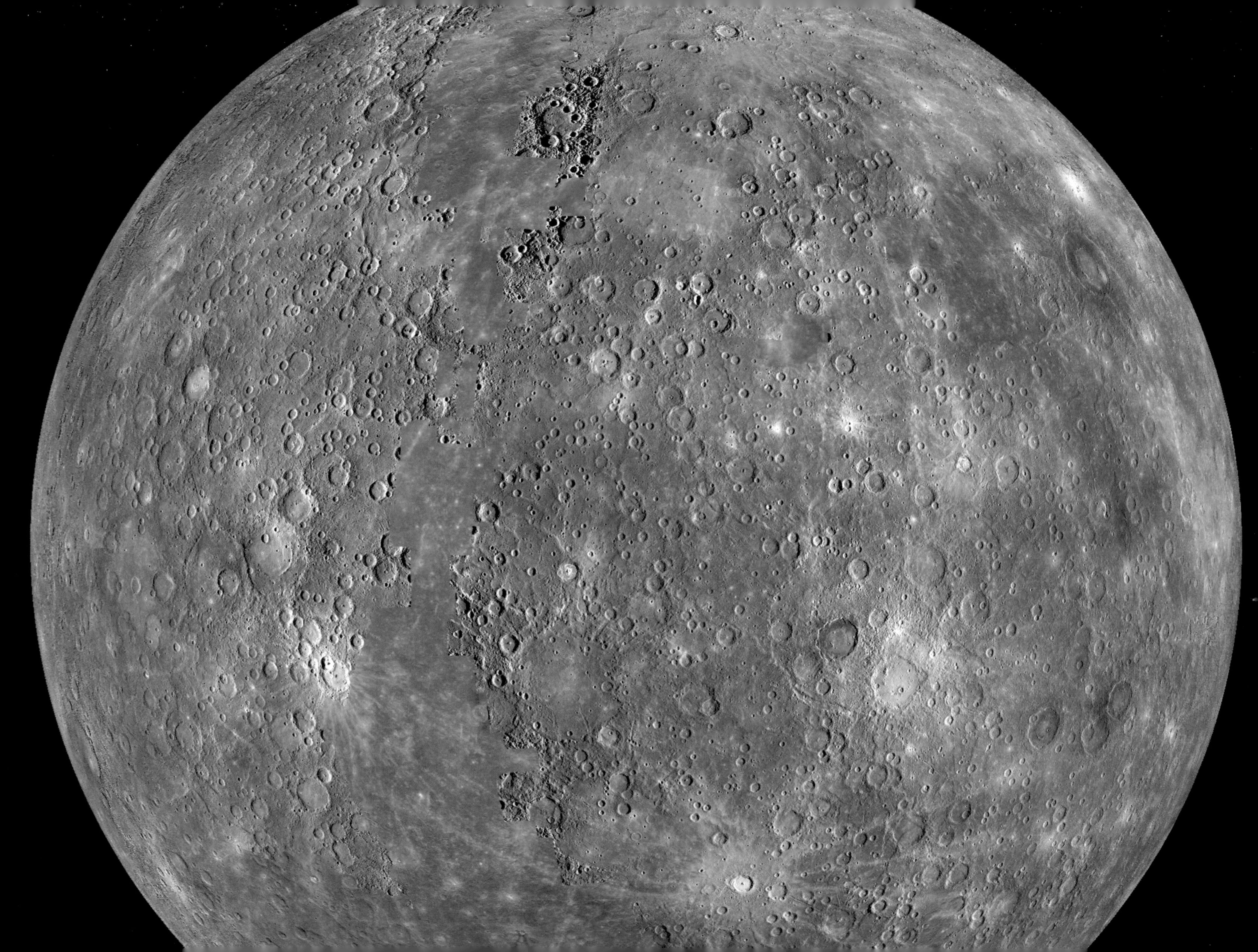

# Mercury & Venus

Mercury and Venus are the inner planets. Closer than Earth to the Sun, they are much more exposed to solar radiation. They are very different worlds to Earth and to each other.

Mercury, named for the fleet-footed messenger of the gods, orbits the Sun in just 88 days at an average distance of only 58 million km (36 million miles). That is 40 per cent of the distance from Earth to the Sun. Its orbit is highly elliptical, carrying it as close as 46 million km (29 million miles) and as far away as 70 million km (43 million miles). Seared by solar glare, Mercury is a baking hot planet devoid of significant atmosphere. At this distance, the Sun's gravity exerts a strong tidal pull, so that Mercury's rotation is locked in resonance with its orbit. It spins on its axis every 59 days, exactly three times for every two orbits of the Sun. On Mercury, the Sun rises and sets only once in every two Mercurian years.

Venus takes its name from the Roman goddess of love and beauty, perhaps a reference to its brilliance in the evening or early morning sky as seen from Earth. Orbiting the Sun every 225 days at an average distance of 108 million km (67 million miles, which is 72 per cent of Earth's distance) it is the planet that comes closest to our own. Because the two planets are similar in size as well as location, Venus has been described as 'Earth's twin'.

In other respects, however, Venus is more like a vision of hell. Its dense, crushing atmosphere of carbon dioxide has created a runaway greenhouse effect. The planet's surface, hidden beneath dense clouds of sulphuric acid, has a temperature of 453°C (847°F), hot enough to melt lead.

Understanding how Venus and Earth became so different is of considerable interest to climate scientists.

OPPOSITE:

**Mercury photographed by *Messenger***

On 18 March 2011, *Messenger* became the first spacecraft to orbit Mercury. Over the following four years it returned over 100,000 images covering 100 per cent of the surface and made several significant discoveries. Because of its proximity to the Sun, inserting a spacecraft into Mercury's orbit is a difficult and complex undertaking, requiring more rocket fuel than getting to Pluto. *Messenger* took nearly seven years to make its journey, during which time it made one fly-by of Earth, two fly-bys of Venus and three fly-bys of Mercury. Each of these planetary encounters provided a gravity assist that adjusted its trajectory to slow the spacecraft down, so it was finally able to enter orbit.

LEFT:

**Mercury transiting the Sun**

With a diameter of 4880km (3032 miles), Mercury is the smallest planet, comparable in size to the largest moons within our Solar System. As seen from Earth, Mercury transits the face of the Sun 13 or 14 times per century. The last such transit was on 11 November 2019 and the next will be on 13 November 2032. In this picture Mercury is a tiny black dot near the centre of the solar disc. The innermost planet is just 1/285th the diameter of the Sun.

RIGHT:

**The first detailed view of Mercury**

On 29 March 1974, *Mariner 10* became the first spacecraft to visit Mercury, passing just 703km (436 miles) above its surface. It went on to make two more fly-bys in the following 12 months, taking over 2800 images covering around 45 per cent of the planet. This was the first time we had seen detail on Mercury, whose surface cannot be resolved with Earth-based telescopes. *Mariner 10* revealed a densely cratered world, similar in overall appearance to Earth's Moon.

Mercury has fewer large-scale features such as plateaux and maria than we see on the Moon but there are highlands, mountains, dorsa (wrinkle ridges), rupes (escarpments) and valleys.

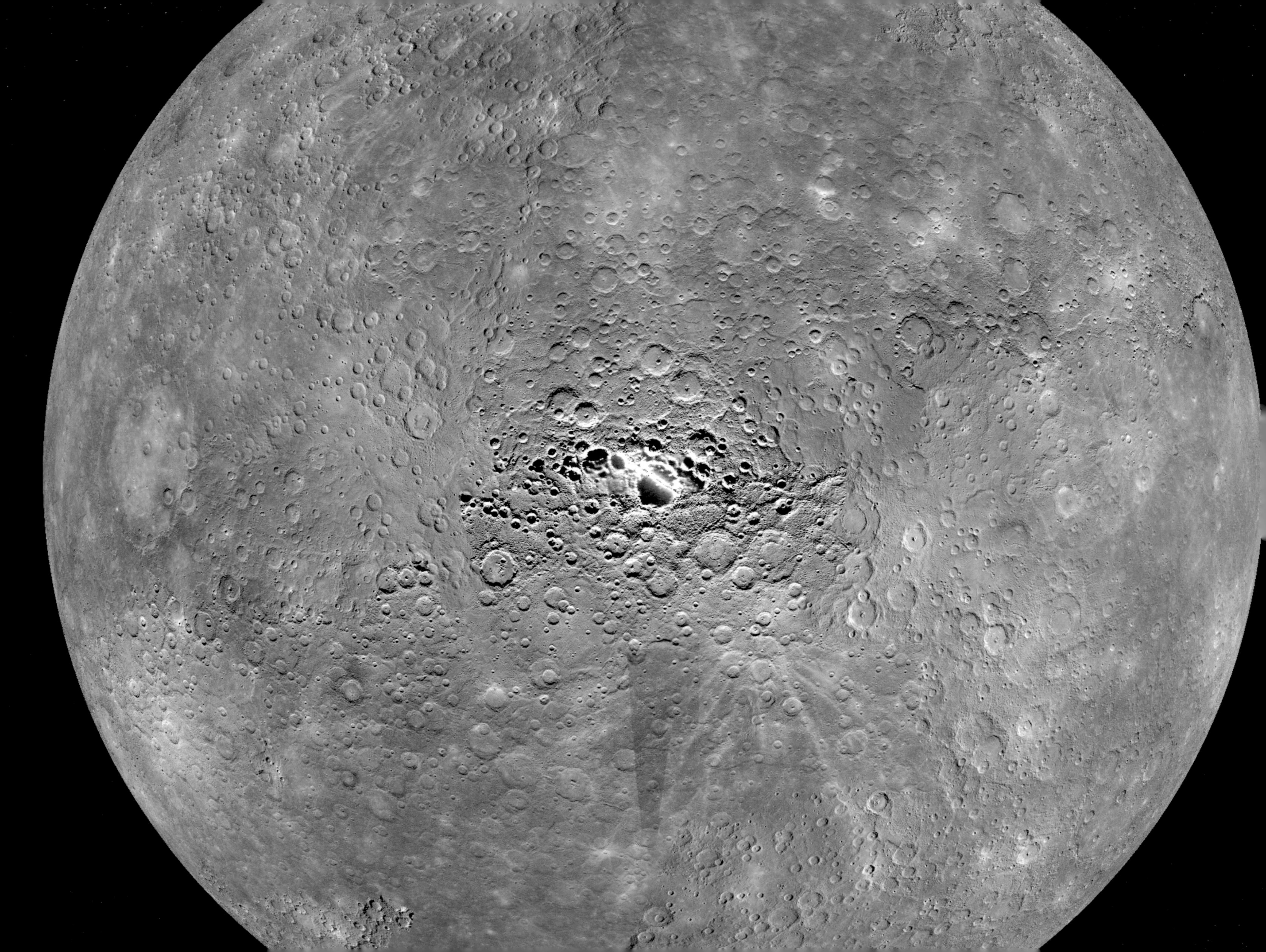

ALL:

**South and north poles of Mercury**

Alone amongst the planets, Mercury has no significant atmosphere. It does, however, have a tenuous exosphere, with a density is so low that the molecules within it rarely collide. The absence of an atmosphere to retain warmth results in extreme differences in temperature between day, when temperatures can reach a sizzling 427°C/800°F (when Mercury is closest to the Sun), and night, which is typically a chilly -173°C (-279°F). This diurnal range of 600°C (1079°F) is more than that experienced on any other planet.

*Messenger*'s orbit enabled it to gain detailed views of Mercury's polar regions. Because the planet's axial tilt is only 2°, the poles are never turned towards the Sun; hence these views are mosaics made up of several constituent images taken as Mercury rotates. The north pole (right) has deep craters in permanent shadow, enabling them to escape the searing heat that bakes most of its surface.

The spacecraft's gamma-ray spectrometer and neutron spectrometer showed that water ice is present in these shaded craters. This finding had been predicted; however, an unexpected discovery was the presence of water vapour in Mercury's exosphere.

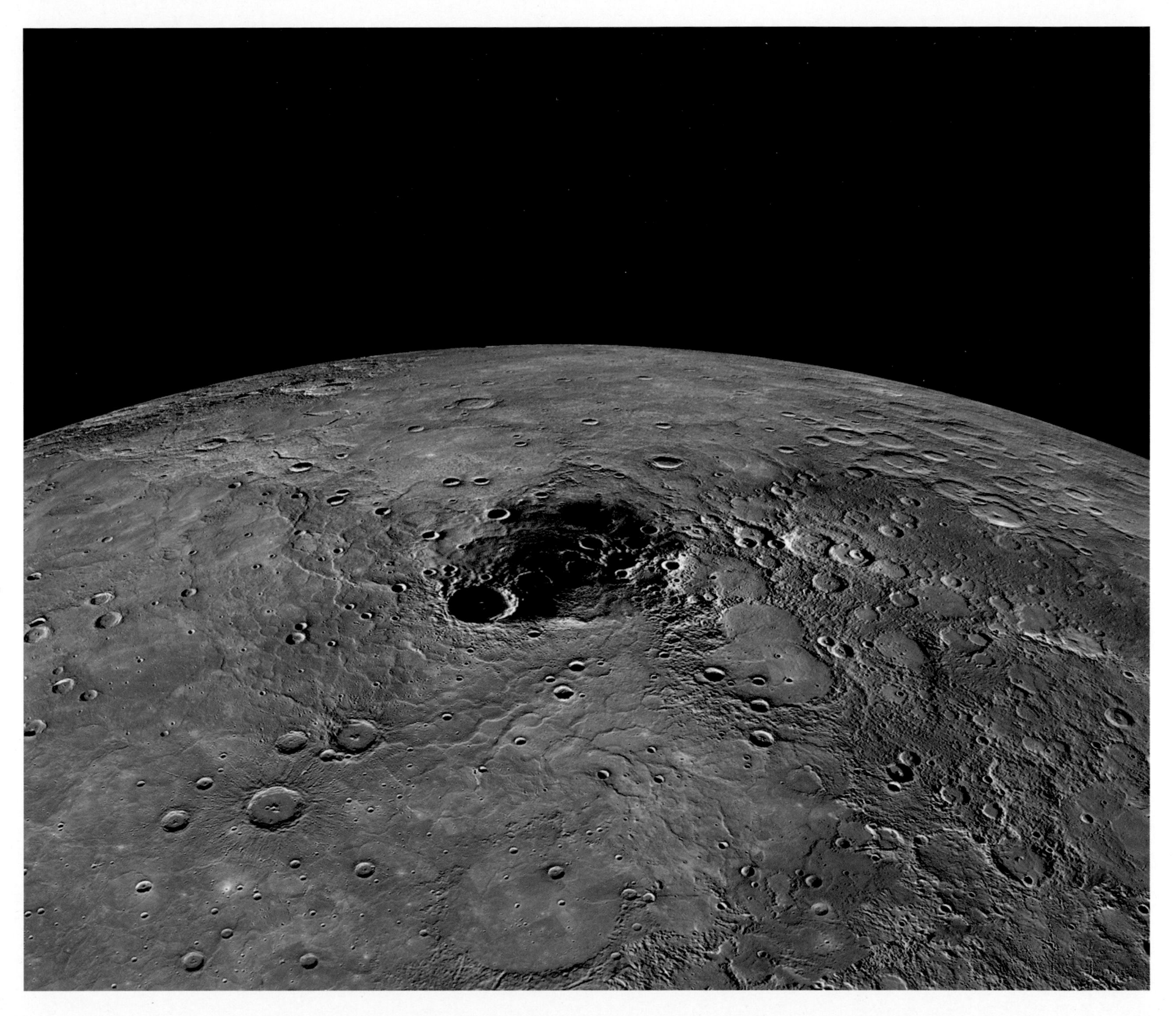

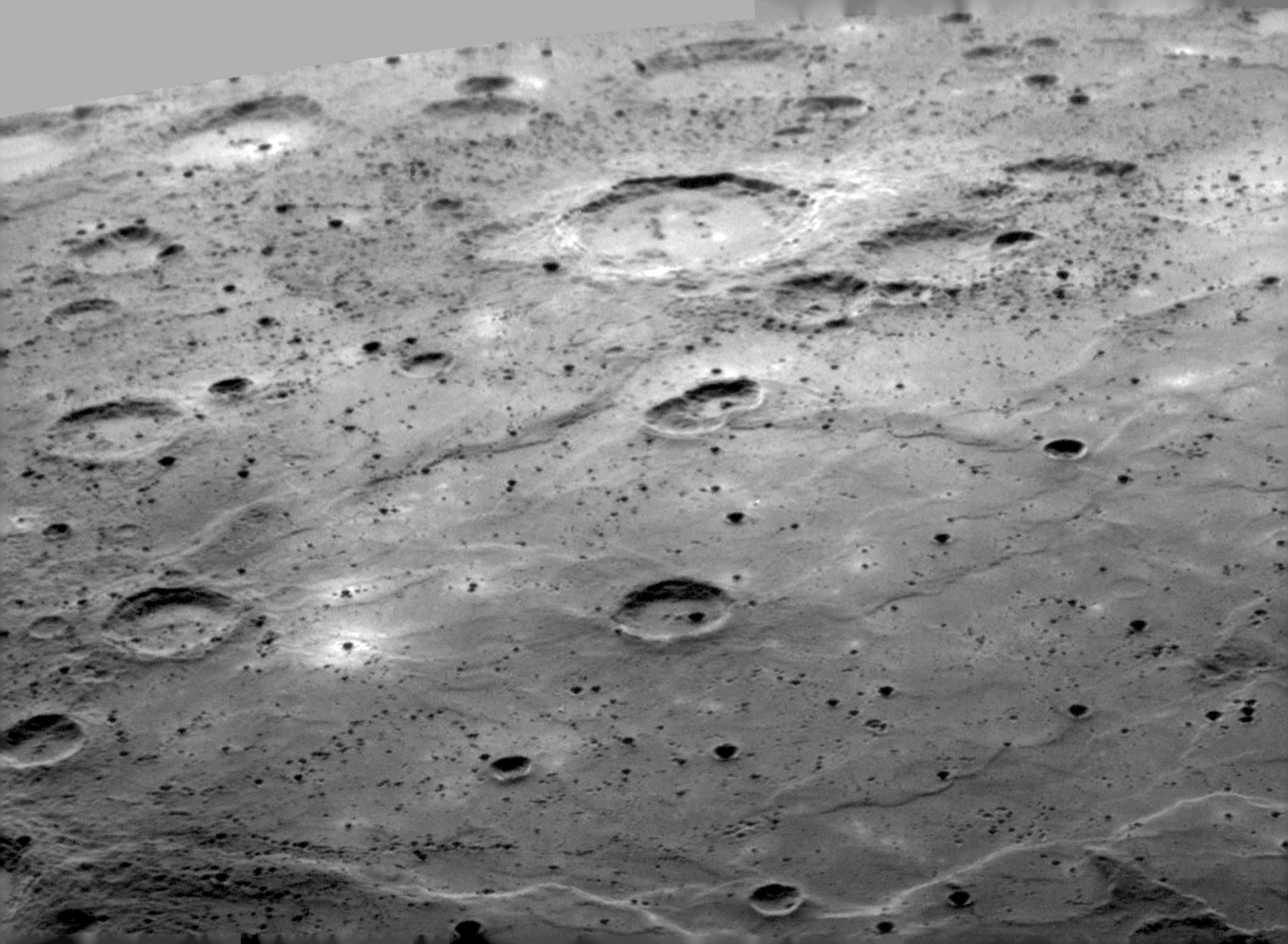

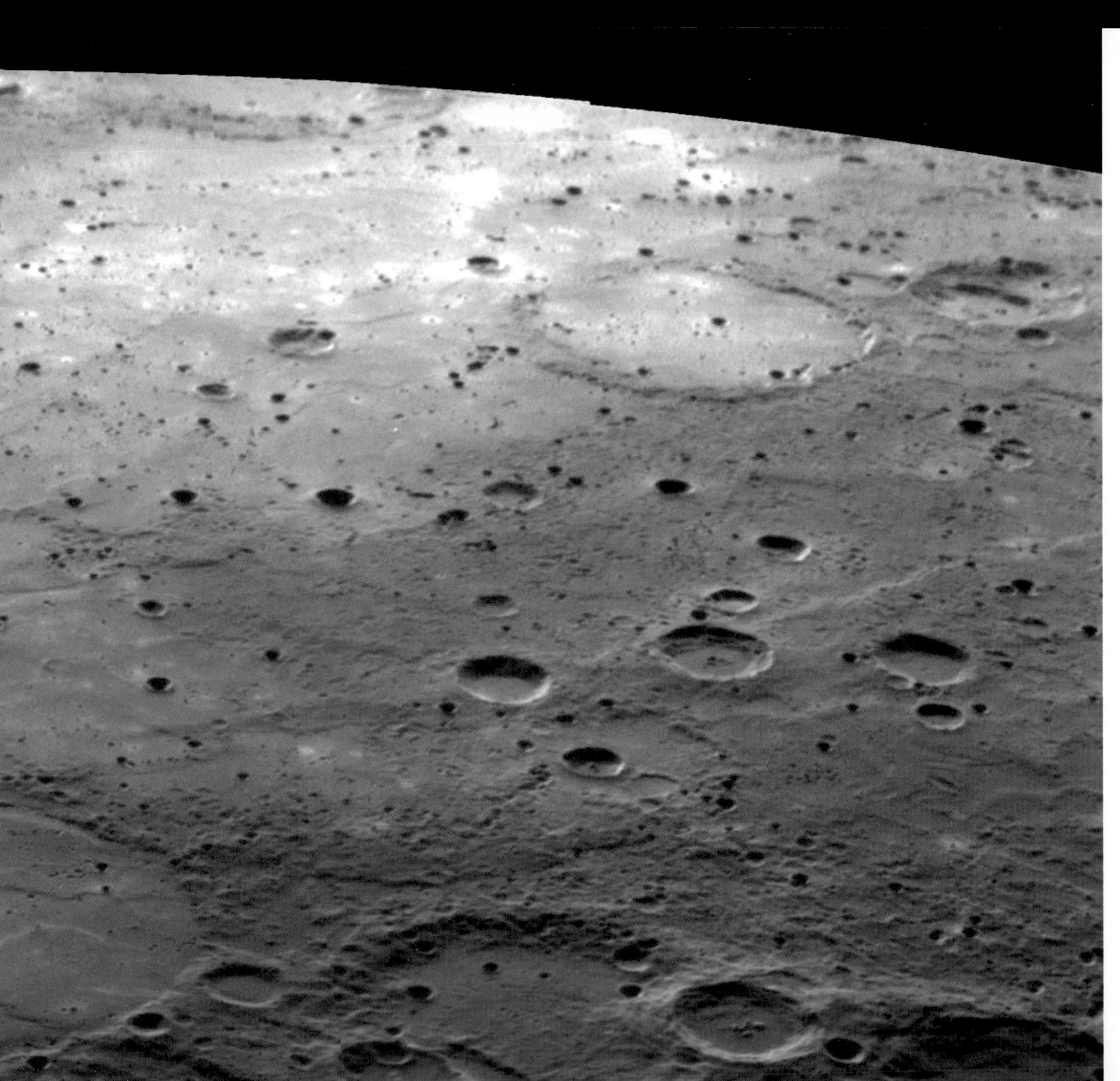

**Volcanoes on Mercury**
*Messenger* gathered evidence of Mercury's internal structure and of past volcanic activity on the surface. This image is a mosaic from several photographs captured during its second fly-by, showing smooth volcanic plains and lava-flooded craters. The spacecraft mapped 51 pyroclastic deposits, resulting from shield volcanoes which appear to have erupted over a prolonged period.

Mercury has the largest core relative to its size of any planet, making it unusually dense. Whilst Earth's core occupies less than one-fifth its volume, over half of Mercury comprises a dense, iron-rich core. It is thought that when newly formed the innermost planet was larger, but its original crust and much of its mantle were stripped away by a cataclysmic impact or by extreme solar heating. *Mariner 10* found that Mercury has a significant magnetic field. *Messenger* further investigated this, leading to the conclusion that Mercury's iron core is still molten.

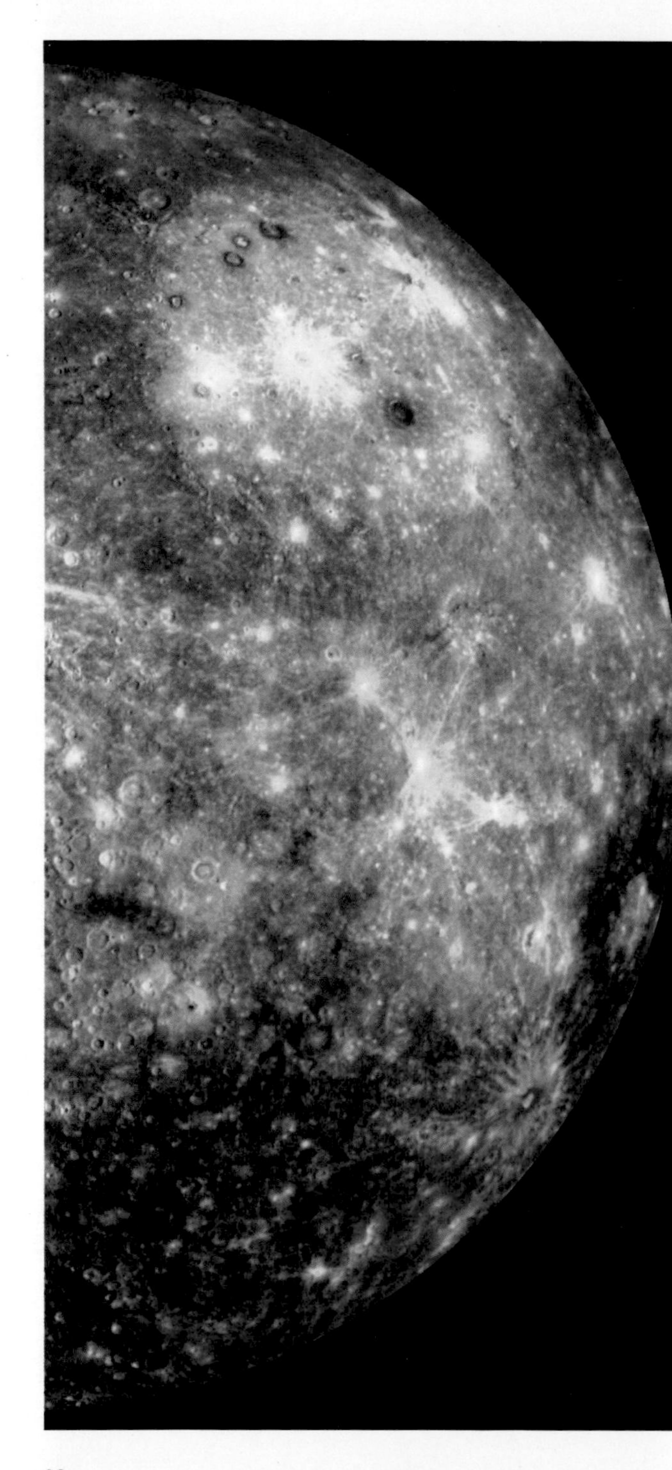

ALL:

**Caloris Basin, Mercury**

Caloris Basin in Mercury's northern hemisphere is an impact crater about 1540km in diameter, making it one of the largest impact basins in the Solar System. The crater is surrounded by a ring of mountains about 2km tall. It was discovered by *Mariner 10* and imaged in its entirety by *Messenger*. These images show the location and detail of the crater, with enhanced colour to show the different rock types. The whole basin was flooded by lava (shown in orange) and subsequently impacted by smaller craters exposing material from below (shown in blue).

It appears that the original impact that created Caloris Basin triggered volcanic activity, flooding the basin with lava to a depth of around 3 km (2 miles). This impact appears to have planet-wide repercussions as there is an area of complex, grooved terrain at the exact antipodal point from the basin. This 'Weird Terrain' is thought to result from seismic shock waves and ejected material travelling around the planet and converging on the opposite side.

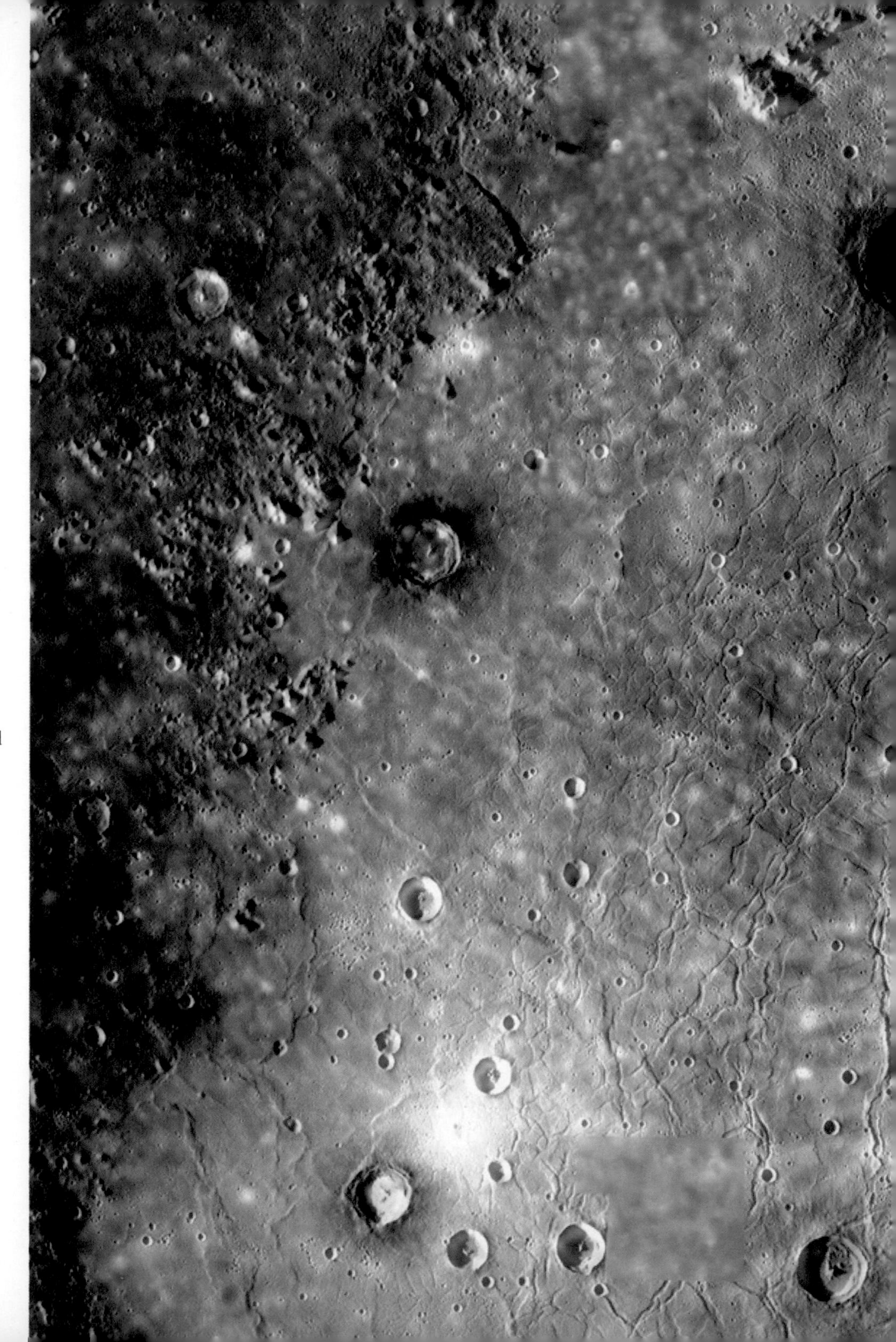

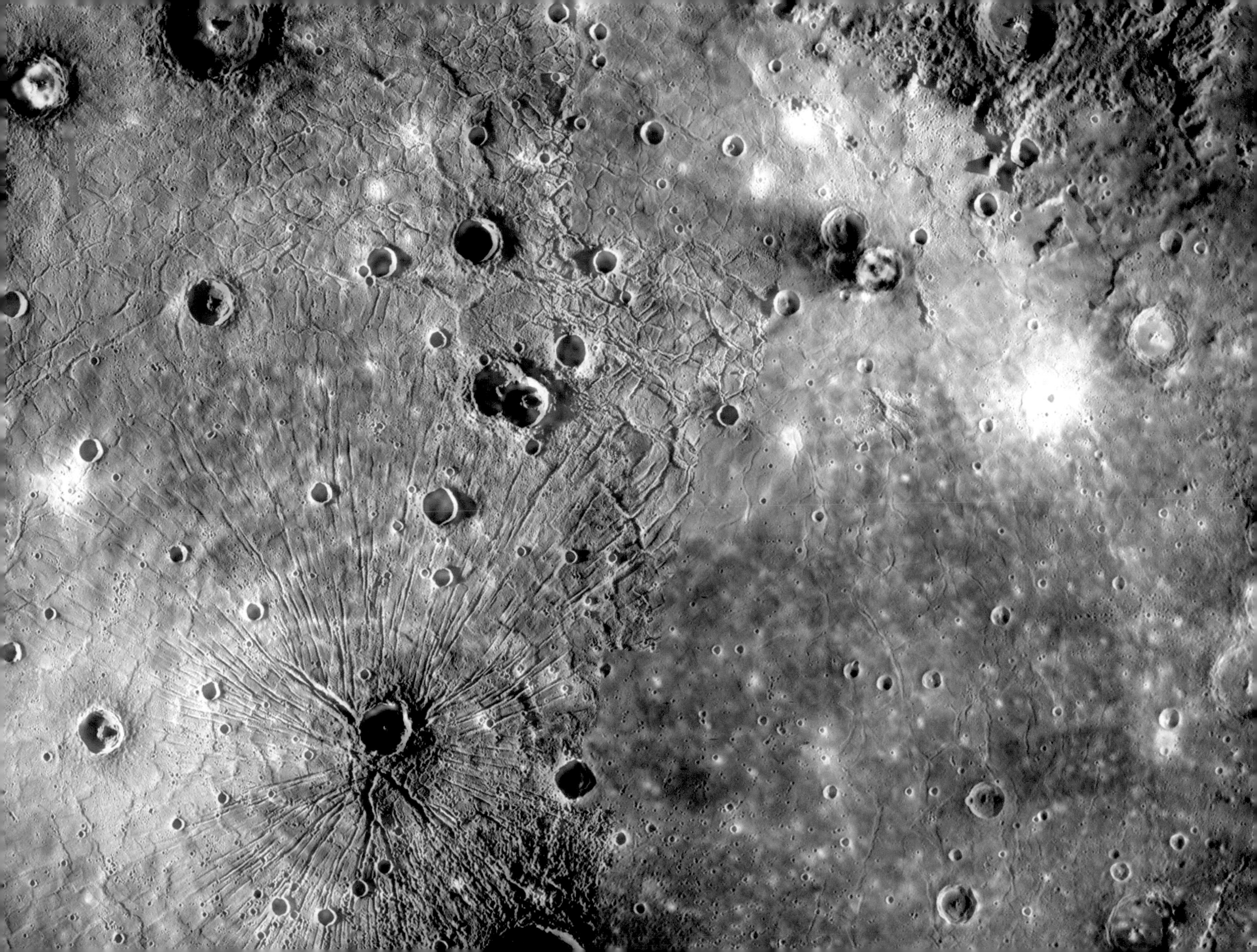

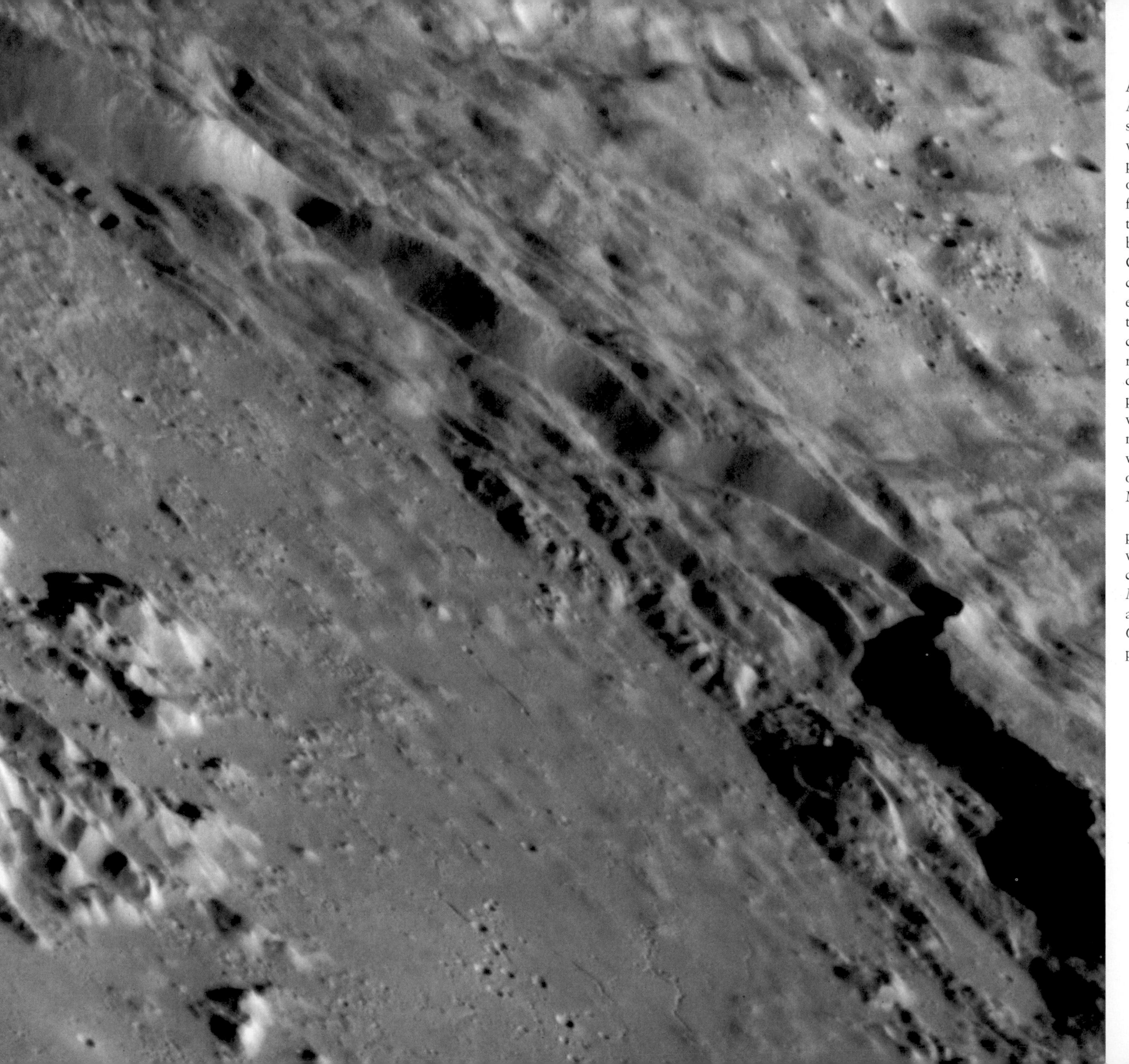

**Abedin crater, Mercury**
Abedin is a complex crater structure complete with wall terraces and a central peak structure, here viewed obliquely by *Messenger*. Its floor is smooth, indicating that it filled with rock melted by the original impact. Chains of smaller secondary craters are the result of ejected material falling back to the surface. Also visible are cracks that formed as molten rock cooled. The shallow depression amidst the central peaks of the crater may be volcanic in origin. Reddish material surrounding this vent is typical of other sites of explosive volcanism across Mercury.

A further mission to the planet named *BepiColombo* was launched in 2018. It comprises two probes, *Mercury Planetary Orbiter* and *Mercury Magnetospheric Orbiter*, due to reach the planet in 2025.

LEFT:

**An early view of Venus**

The first spacecraft to visit Venus was *Mariner 2* in 1962; however, it did not carry a camera. This image was returned by *Venus Pioneer* in 1979, showing the dense clouds that completely obscure views of the surface in visible light. Four probes carried by *Venus Pioneer* entered the atmosphere and found it comprises 96 per cent carbon dioxide, with a surface pressure 93 times that of Earth (equivalent to the pressure at a depth of 900m/3000ft in Earth's oceans). This suffocating atmosphere makes Venus the hottest planet in the Solar System at 453°C (847°F).

RIGHT:

**Venus revealed**

Radar enables us to penetrate the clouds of Venus and map its surface. This was the primary purpose of the *Magellan* mission, launched aboard the space shuttle in May 1989. *Magellan* spent four years in orbit around Venus from 1990 to 1994 completing six mapping cycles. This global view of Venus is a compilation of radar mosaics from *Magellan* and *Venus Pioneer* orbiter data. Since radar tells us nothing about colour, the simulated hues are based on visible light images from the surface recorded by *Venera 13* and *Venera 14*.

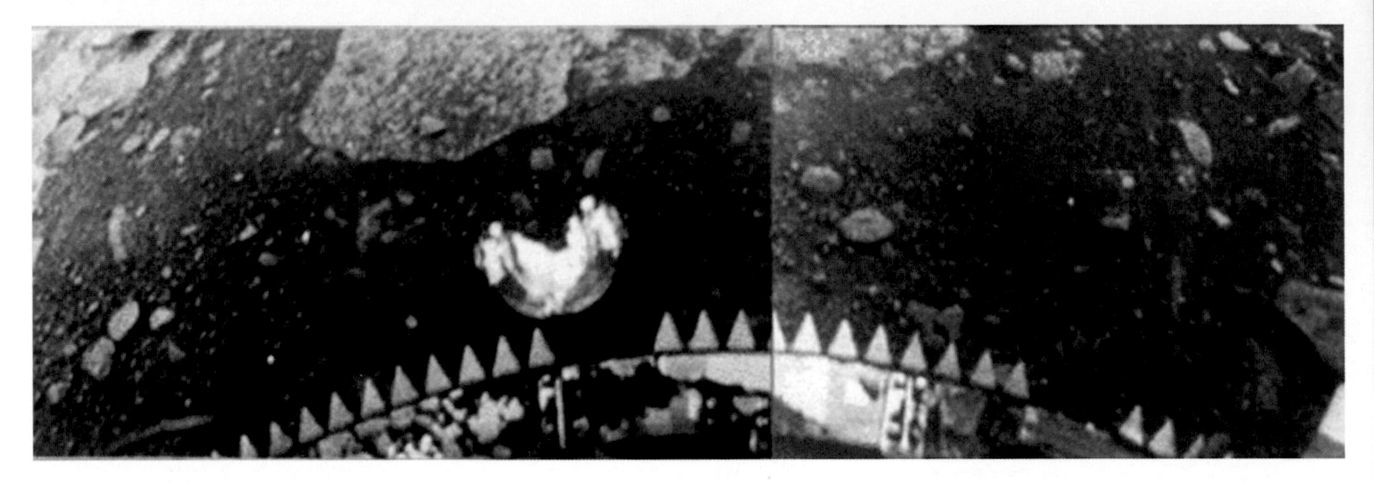

ABOVE:

**Landing on Venus**

Between 1970 and 1984, a series of *Venera* probes succeeded in landing on Venus. Early attempts were crushed by atmospheric pressure and failed to reach the ground. In December 1970, *Venera 7* became the first spacecraft to survive the descent and transmit from the surface. Owing to the extremely hostile conditions, the probe survived just 23 minutes after landing. Five years later, *Venera 10* returned the first images from the surface, which were in monochrome.

This colour image was taken by *Venera 13* in 1982, showing part of the spacecraft and flat rocks amongst which it landed. During just 127 minutes of operation in temperatures of 457°C (854°F), the lander also recorded the first sounds from another planet (including Venusian wind), dug a soil sample and used a spectrometer to identify the presence of leucite basalt.

Venus orbits the Sun every 225 days and, in contrast to most other planets, it rotates in the opposite direction to its orbit. This retrograde rotation takes 243 days to complete, the slowest of any planet in our Solar System. Consequently a Venusian day lasts longer than a Venusian year. An observer on the surface cannot see the Sun due to thick clouds but, if it were possible to observe it, the Sun would rise in the west and set in the east.

RIGHT:

**Maat Mons**

A computer-simulated terrain model (with vertical scale exaggerated) of Maat Mons. Rising 8 km (5 miles) from the average surface elevation, this massive shield volcano is the highest on Venus.

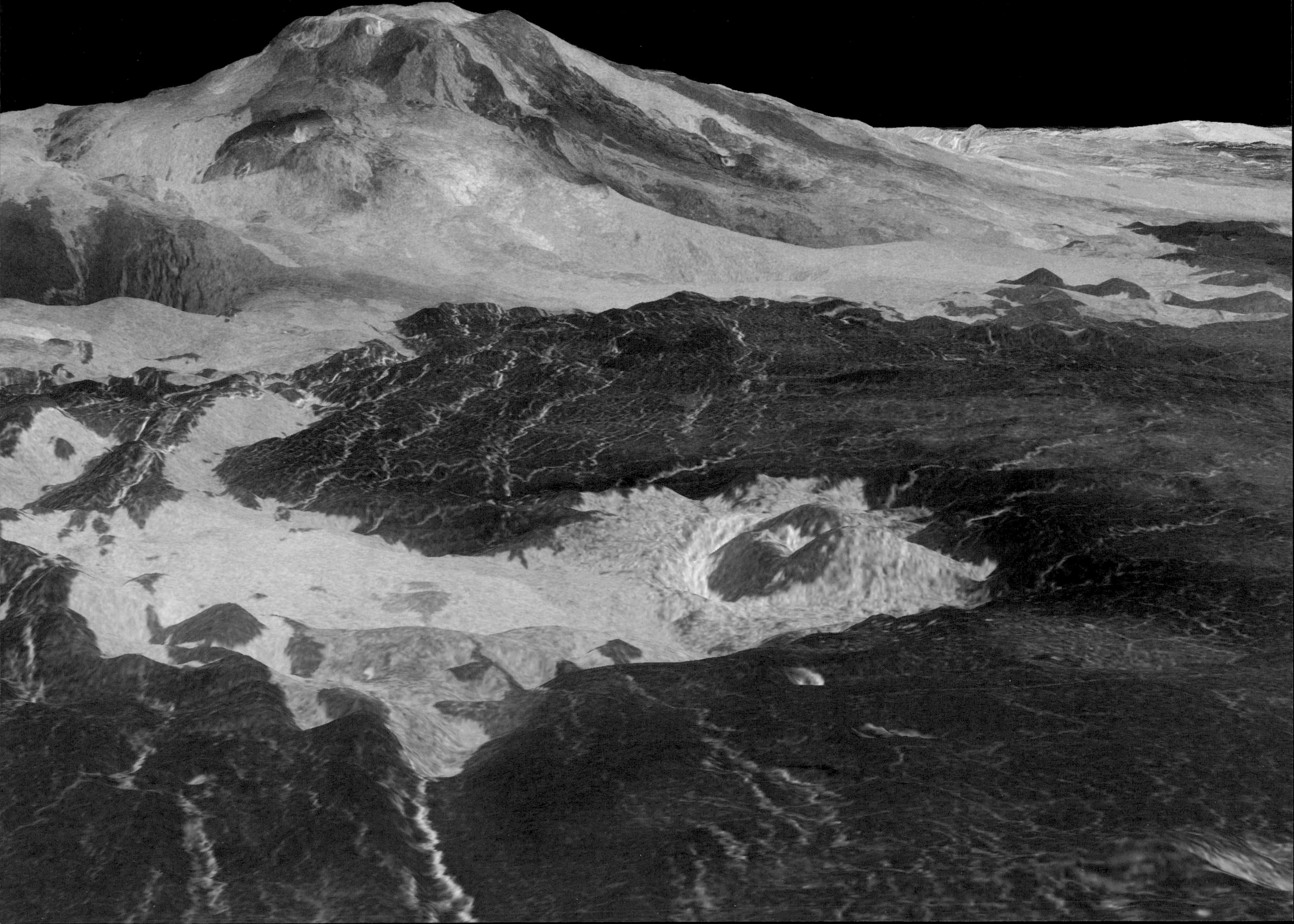

**Venusian volcanoes**
Venus has more volcanoes than any other planet, including 167 in excess of 100km (60 miles) wide (comparable to the largest volcano on Earth). Many are shield volcanoes whilst others are 'pancake domes', thought to be formed when viscous lava erupts under Venus's high atmospheric pressure. Radar images from *Magellan* show three unusual volcanoes in Guinevere Planitia, all circular with steep sides, the largest of which is about 50km (30 miles) in diameter. Most Venusian volcanoes are likely to be extinct but there is evidence of volcanism within the last few million years. *Venus Express* observed four transient infrared hotspots that may indicate current activity. Levels of atmospheric sulphur dioxide have been found to vary by a factor of 10 between 1978 and 2006, suggesting that volcanic eruptions may have occurred during that time.

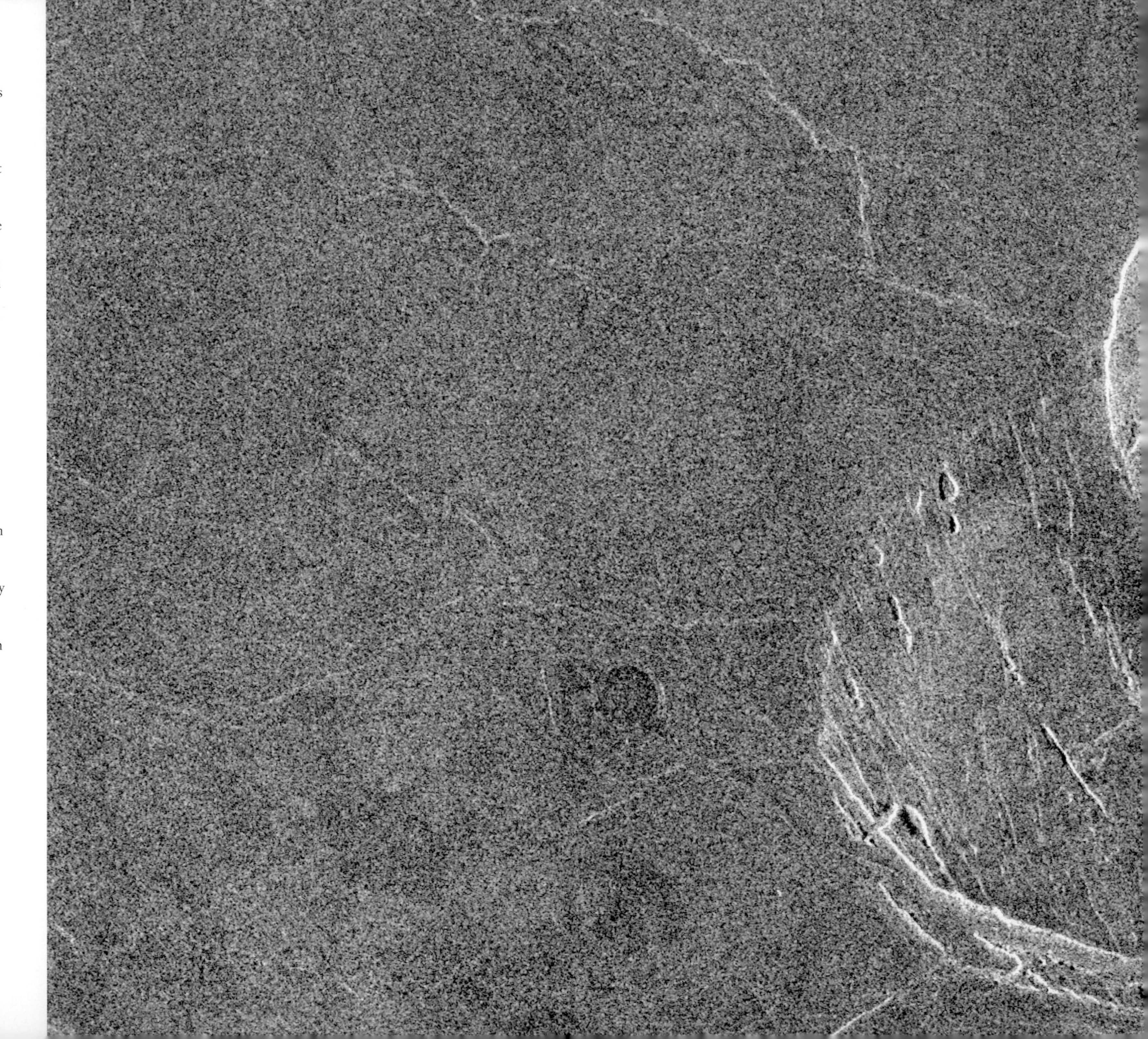

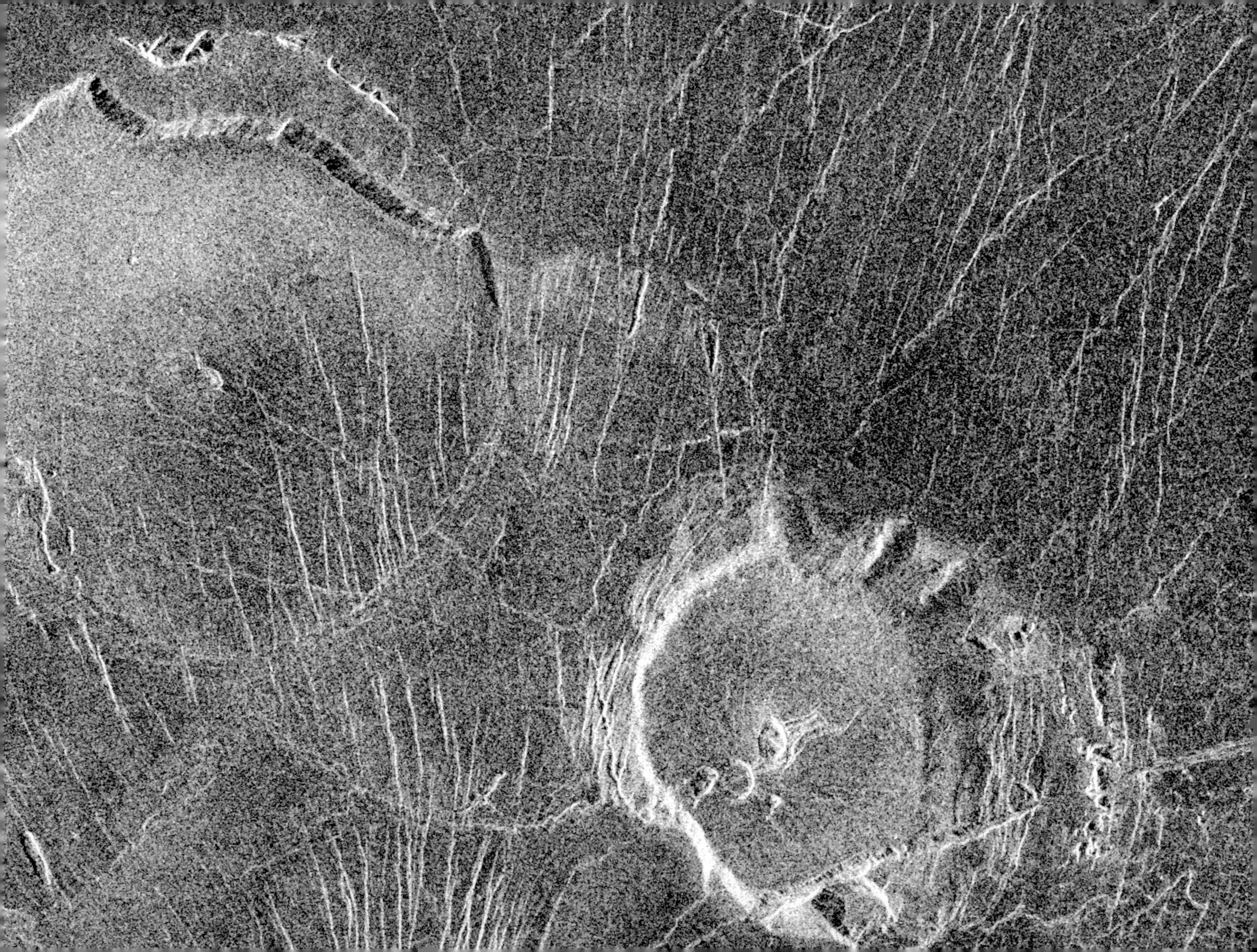

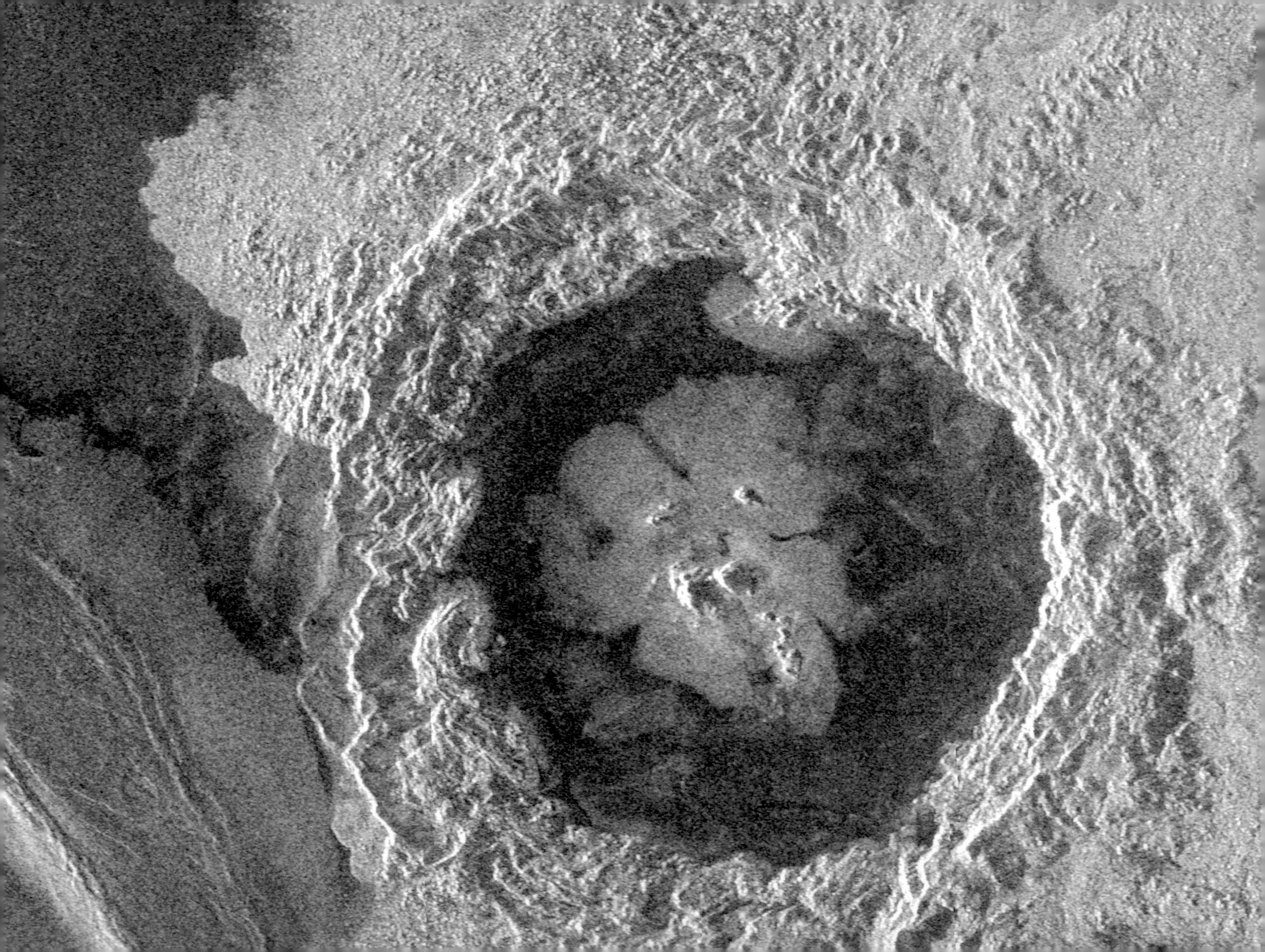

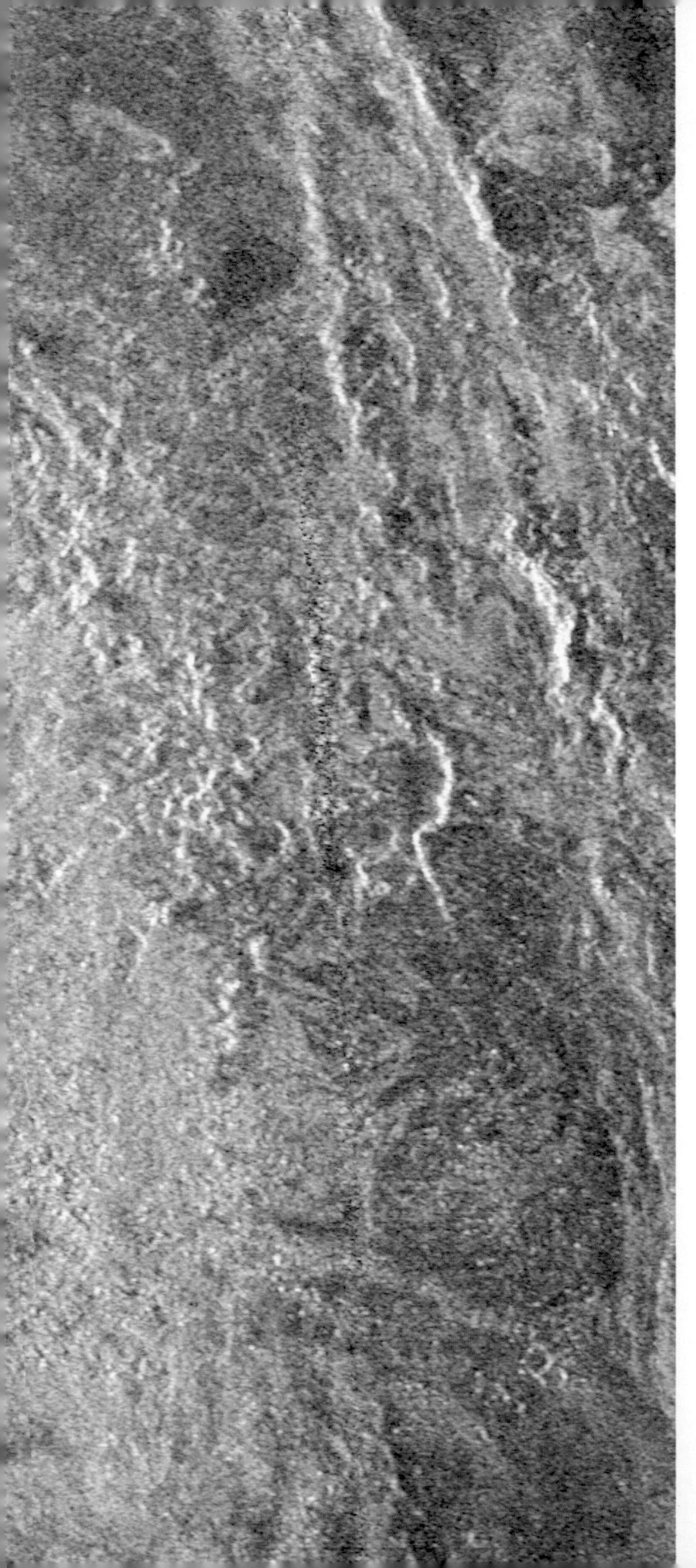

LEFT:

**Venusian craters**

Almost 1000 impact craters are scattered across Venus. None is smaller than 3km in diameter because incoming meteoroids less than about 50m (160ft) in diameter burn up or fragment in their passage through the thick atmosphere. This image shows Dickinson, a complex impact crater 69km (43 miles) in diameter. There is a partial central ring and surroundings of rough-textured material thrown out by the impact. The crater's smooth floor results from flooding by molten rock, either melted by impact or resulting from volcanic activity triggered by the impact.

RIGHT:

**Venusian channels**

The surface of Venus contains over 200 channel systems (valles), superficially resembling rivers. They are believed to have been formed not by flowing water but by molten lava.

Baltis Vallis is the longest valle so far discovered, 1.8km (1.1 miles) wide and more than 7000km (4300 miles) in length. Visible features include meanders and oxbows similar to those of terrestrial rivers; however, tributaries have not been observed. The oldest valles are crossed by fractures and wrinkle ridges; curiously they appear to run uphill as well as downhill. This suggests that valles are old and the surface has been warped by tectonic activity after they were formed.

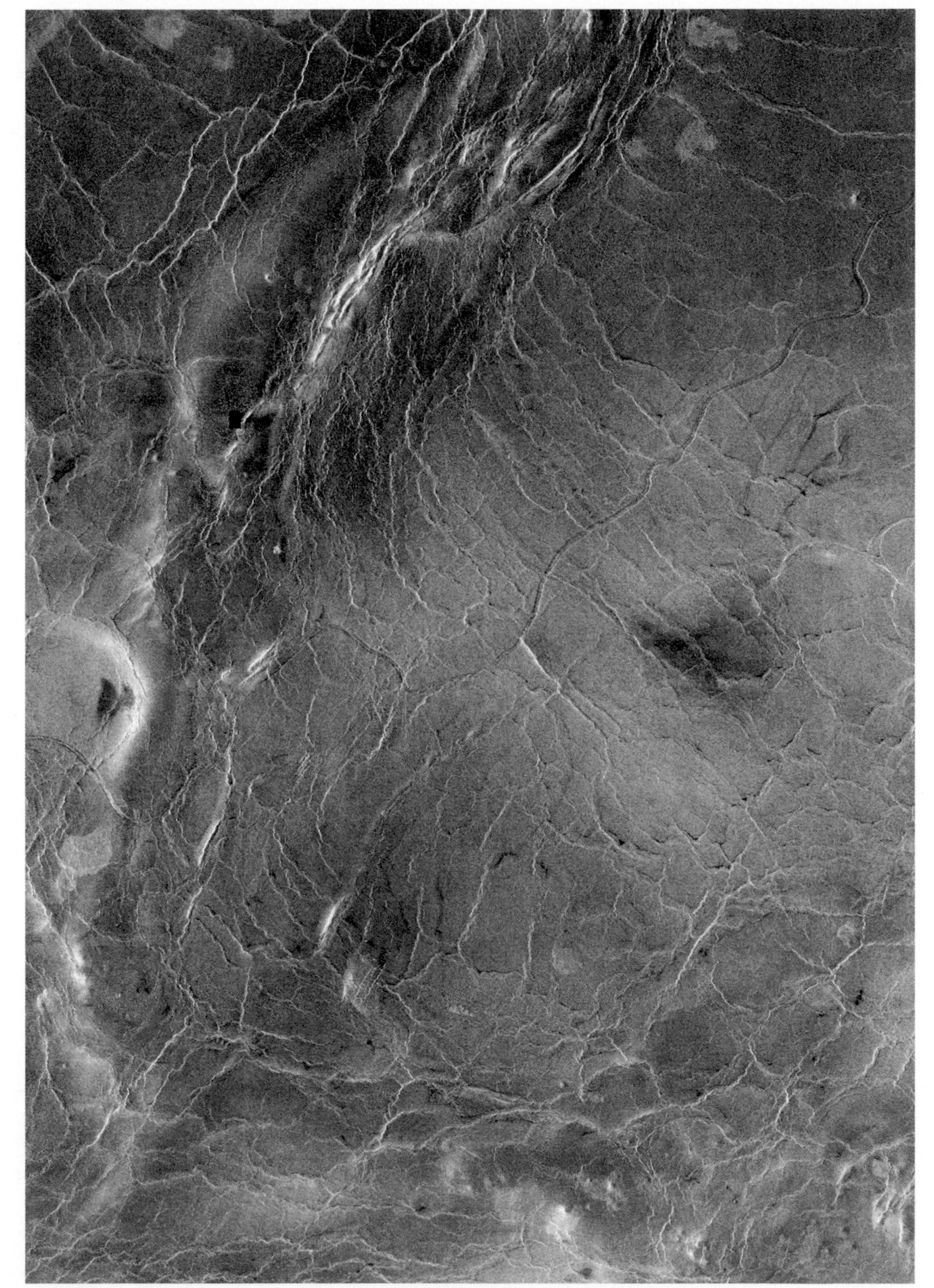

RIGHT:

**Venusian clouds**

Some 50km (30 miles) above the surface are thick clouds, composed mainly of sulphuric acid droplets. This infrared image from the *Akatsuki* orbiter, which arrived in 2015, shows structure within them. The clouds reflect three-quarters of sunlight back to space, making Venus the brightest planet in Earth's night sky. The intensity of sunlight reaching the Venusian ground is similar to a partly cloudy day on Earth. In fact, thanks to continuous cloud cover, the surface receives less solar energy than Earth, despite Venus being closer to the Sun. Whilst wind speeds on the surface of Venus are low, those of the upper atmosphere reach 360km/h (220mph).

The Venusian surface is extremely toxic to life. It has been suggested that conditions in the atmosphere above the clouds, where pressure and temperature are similar to those found on Earth's surface, could be more hospitable. In 2020 it was reported that phosphine had been detected, leading to speculation that biological processes could be responsible. However, later research has suggested that the spectroscopic signal attributed to phosphine may have been mistaken.

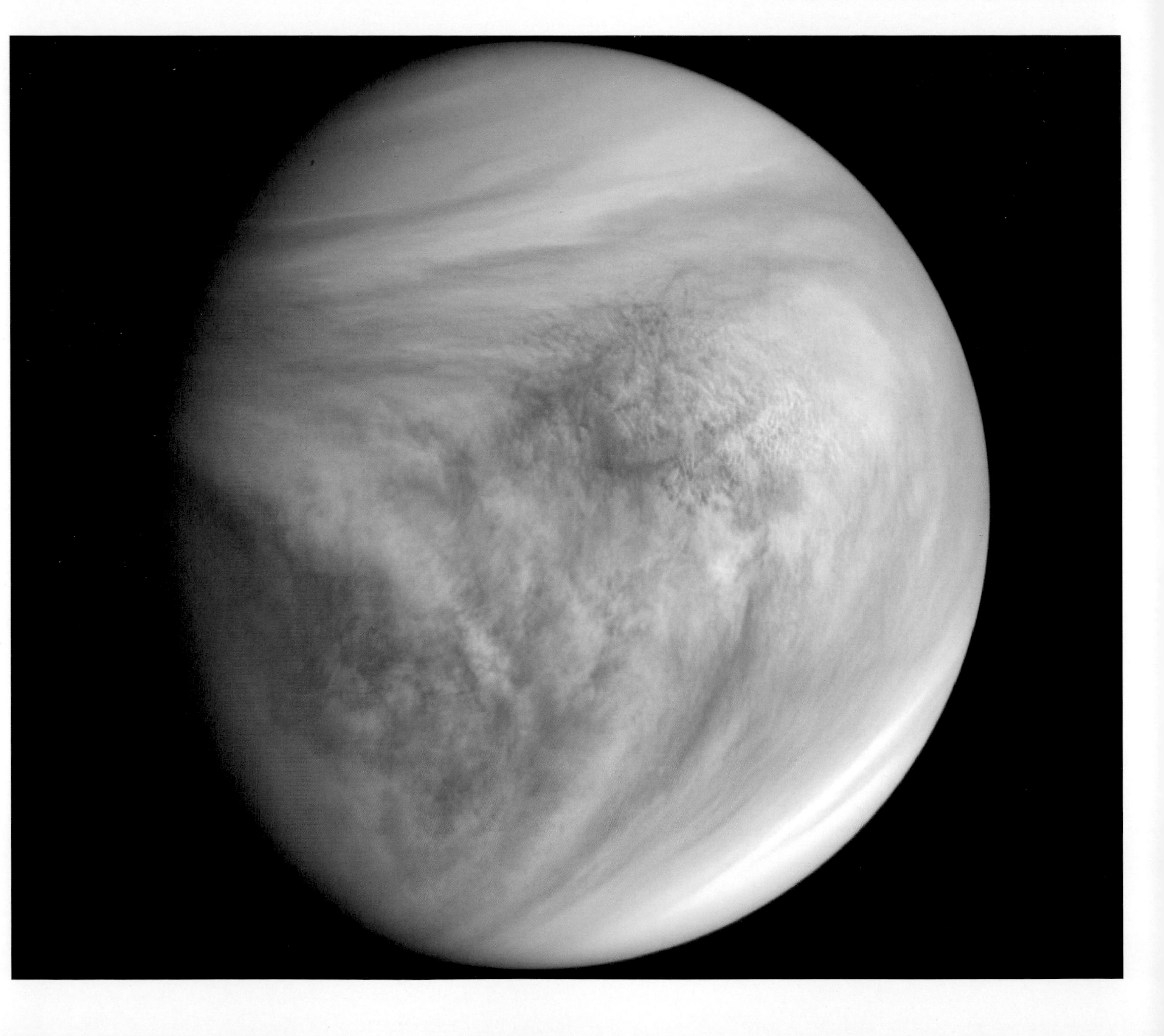

RIGHT:

**Venus Express**

Seen in this artist's impression, *Venus Express* spent over eight years in orbit from 2006 to 2014 with the primary purpose of studying atmospheric dynamics. Evidence has been gathered that early on its history, Venus may have had much lower atmospheric pressure and possibly oceans of liquid water. As the Sun heated up over hundreds of millions of years, this water evaporated. Without water to lubricate subduction zones, plate tectonics (if they ever existed) ground to a halt. This in turn removed a major long-term geological carbon sink, contributing to the build-up of carbon dioxide in the Venusian atmosphere. As a potent greenhouse gas, carbon dioxide led to catastrophic planetary warming. According to this theory, these are key reasons why Venus and Earth have evolved from similar beginnings to the dramatically different worlds we see today.

# Earth & Moon

Around eight billion humans currently live on Earth but, throughout the whole of human history, only 24 people have seen the entire planet with their own eyes. However, photographs taken by *Apollo* astronauts are familiar throughout the world. They are credited with fostering global environmental consciousness, as we saw for the first time the fragility of Earth's life-supporting ecosystems set against a backdrop of vast, black, empty space. One of the most striking features is the predominantly blue colour of our planet, 71 per cent of which is covered by water.

Water is essential to all known forms of life. The fact that Earth, alone amongst the planets, has suitable temperature and pressure for liquid water to exist on its surface arises from a serendipitous combination of circumstances. Our planet is located at just the right distance from the Sun – like Goldilocks, our position is neither too hot nor too cold. Our atmosphere creates just the right amount of greenhouse warming to keep temperatures suitable for life. Earth is a geologically active planet with volcanoes, earthquakes and plate tectonics. In recent decades we have come to understand how these too are integral to the stability of Earth's climate. Our planet has a molten core of iron, which generates a global magnetic field, protecting us from solar radiation.

Earth is the largest of the four 'terrestrial' planets and the only one to have a large natural satellite. The Moon is one-quarter of Earth's diameter, two per cent of its volume and 1/81$^{st}$ of its mass. No other planet has a moon more than 1/4000$^{th}$ of its own size. The Moon has major effects on Earth, including creating tides in our oceans, slowing our planet's rotation over geological time and stabilizing its axial tilt close to its present value of 23.4°. Indeed, these effects are so important that the Moon has played a key role in maintaining the climatic stability of our planet, its long-term habitability and the evolution of terrestrial life.

OPPOSITE:

**Blue marble**

The almost fully illuminated Earth photographed by the crew of *Apollo 17* is one of the most reproduced images in history. It was taken on 7 December 1972 from a distance of 29,000km (18,000 miles), five hours after launch. Africa, Madagascar and the Arabian peninsula can be seen in their entirety. Two weeks before the December solstice, Antarctica was fully illuminated.

OPPOSITE:

**Ocean**

Water covers 71 per cent of our planet's surface to an average depth of 4km (2.5 miles). The total amount of water on our planet is estimated to be 1.36 billion cubic km, of which 97 per cent is salt water and a further 1.7 per cent is ice. Surface fresh water such as lakes, reservoirs and rivers make up just 0.3 per cent of the total volume.

We often speak of different oceans such as the Pacific, Indian and Atlantic but in fact Earth has a single interconnected ocean. It is crucial to regulating climate both globally and locally. Water retains heat longer than land so the ocean moderates extremes of temperature, whilst water evaporated from the ocean falls as rain over land.

The origin of all this water is debated. When Earth was newly formed four and a half billion years ago, it was too hot for water to condense. Water would have been present at least three times further from the Sun, within and beyond the asteroid belt. It may be that water arrived on the primordial Earth as a result of bombardment by asteroids (which have similar isotopic ratios to our ocean) or comets. Geological evidence from ancient underwater volcanic eruptions indicate that water was present on Earth by 3.8 billion years ago.

RIGHT:

**Polar ice caps**

The Arctic and Antarctic ice sheets cover around five per cent of Earth's surface and lock up some 30 million cubic km of frozen water (sufficient to raise global sea levels by around 70m/220ft). Over geological time, this has not always been the case. We are currently living in an ice age, during which the extent of polar ice periodically increases (glacials) and retreats (interglacials) as a result of 'wobbles' in the eccentricity of Earth's orbit and its axial tilt, termed Milankovitch cycles. Our present interglacial started around 10,000 years ago.

It is thought that conditions for the ice age may have been triggered by the position of the continents, with Antarctica positioned over the south pole and an almost land-locked ocean around the north pole, and by the uplift of the Himalayas. Studies of ice cores from Antarctica show that over the last 800,000 years there has been a close correlation between global temperature and levels of carbon dioxide ($CO_2$) in the atmosphere.

## Plate tectonics

Earth's continents have not always occupied their present positions. In fact they are in constant motion at speeds of a few centimetres per year. The reason is that Earth's crust is divided into eight major (and many minor) plates. The crust is typically five to 30km (three to 18 miles) thick and below it is the mantle, which makes up 84 per cent of Earth's volume. The mantle is solid rock but it is hot, with the result that over long timescales it behaves like a viscous fluid. Plumes of heat rise through the mantle by convection, causing the crustal plates to move, like icebergs drifting in ocean currents.

Plates are pushed apart at mid-ocean ridges, where new oceanic crust is formed. The most accessible of these is Silfra in Iceland (pictured right), where a rift has formed between the diverging North American and Eurasian plates. The Great Rift Valley in Africa (pictured opposite), over 6000km (3700 miles) in length with its string of lakes, is a divergent plate boundary which over the next 10 million years will lead to the separation of what is now East Africa from the rest of the continent.

Elsewhere, plates are being pushed together. Subduction occurs as one plate is forced beneath the other and sinks into the mantle. This is the case around the rim of the Pacific Ocean, where subduction is associated with powerful earthquakes.

Earth is the only planet to have active plate tectonics, driven by heat from its core. One reason for this is thought to be the presence of water in our oceans acting as a lubricant, enabling plates to slide past and under each other.

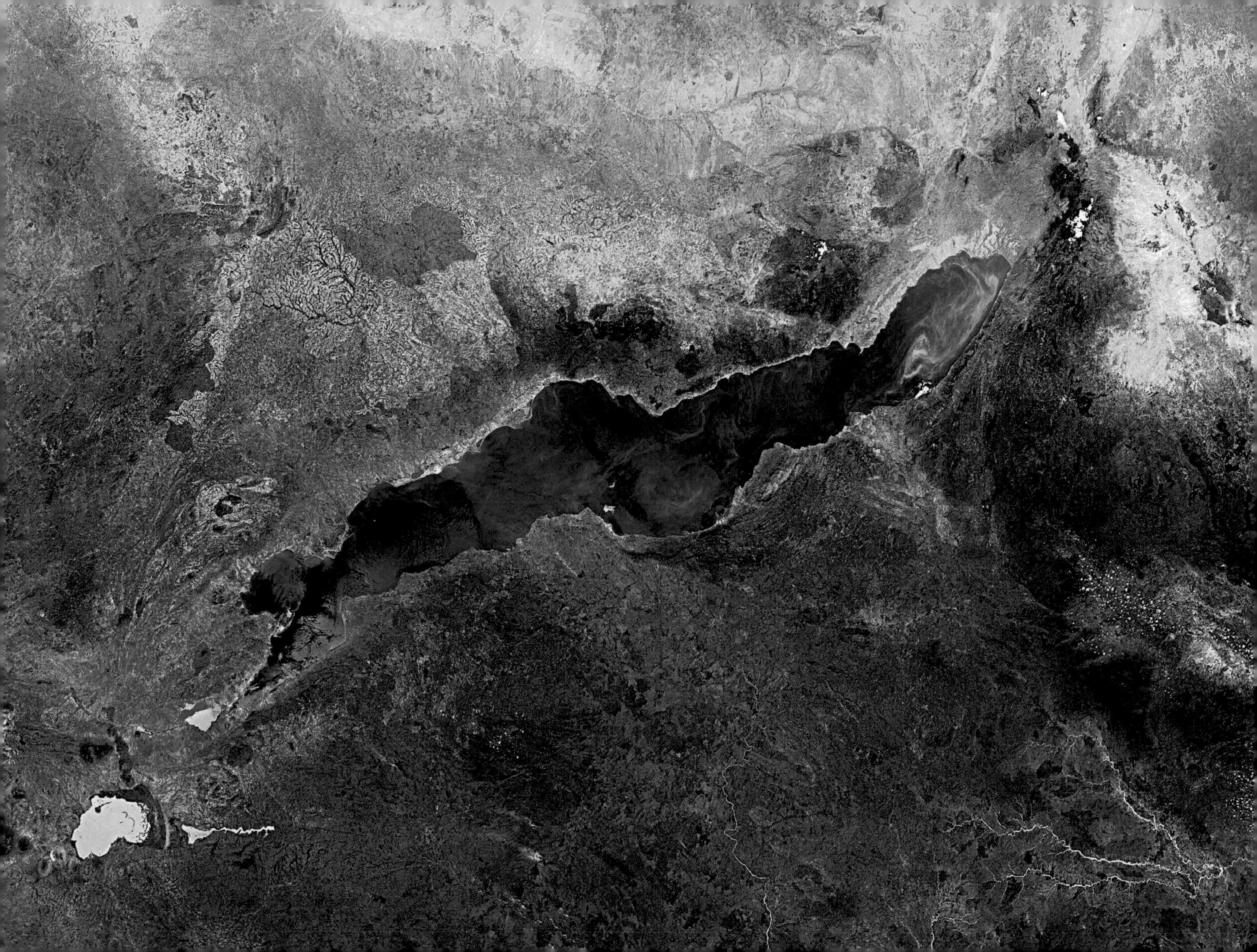

## Mountains

Earth's major mountain ranges are the result of plate tectonics in action. Our highest mountains are the Himalayas, formed over the last 50 million years as the Indian Plate collided with the Eurasian Plate, causing the crust to buckle upwards. The highest peak, Mount Everest, is 8849m (29,032ft) above sea level and 19,833m (65,068ft) above the bottom of the Mariana Trench, which is the deepest point of the ocean floor.

The world's longest mountain range is the Andes (pictured left), formed along the western rim of South America as the continent collides with the Nazca Plate (underlying part of the Pacific Ocean). Because oceanic crust is denser than continental crust, the Nazca Plate is being subducted beneath South America.

LEFT:
**Geysers**
Geysers are hot springs that erupt intermittently, ejecting water and steam. The erupting water is heated underground in proximity to magma until it boils under pressure. Specific hydrological and volcanic conditions are needed to produce geysers, so they are quite rare.

Major geyser fields are scattered across Earth, particularly in Yellowstone National Park (USA), Strokkur in Iceland (pictured left), Kamchatka in Russia, Chile and New Zealand.

OPPOSITE:
**Volcanoes**
Earth has around 1350 potentially active volcanoes. These ruptures in the crust allow molten rock (lava), ash and gas to escape from a magma chamber below the surface. Most volcanoes occur at the boundaries of tectonic plates, for example around the rim of the Pacific, referred to as a 'Ring of Fire'. One of the world's most active volcanoes is Stromboli (pictured right), which has erupted almost continually for over 2000 years as a result of the African and Eurasian plates colliding.

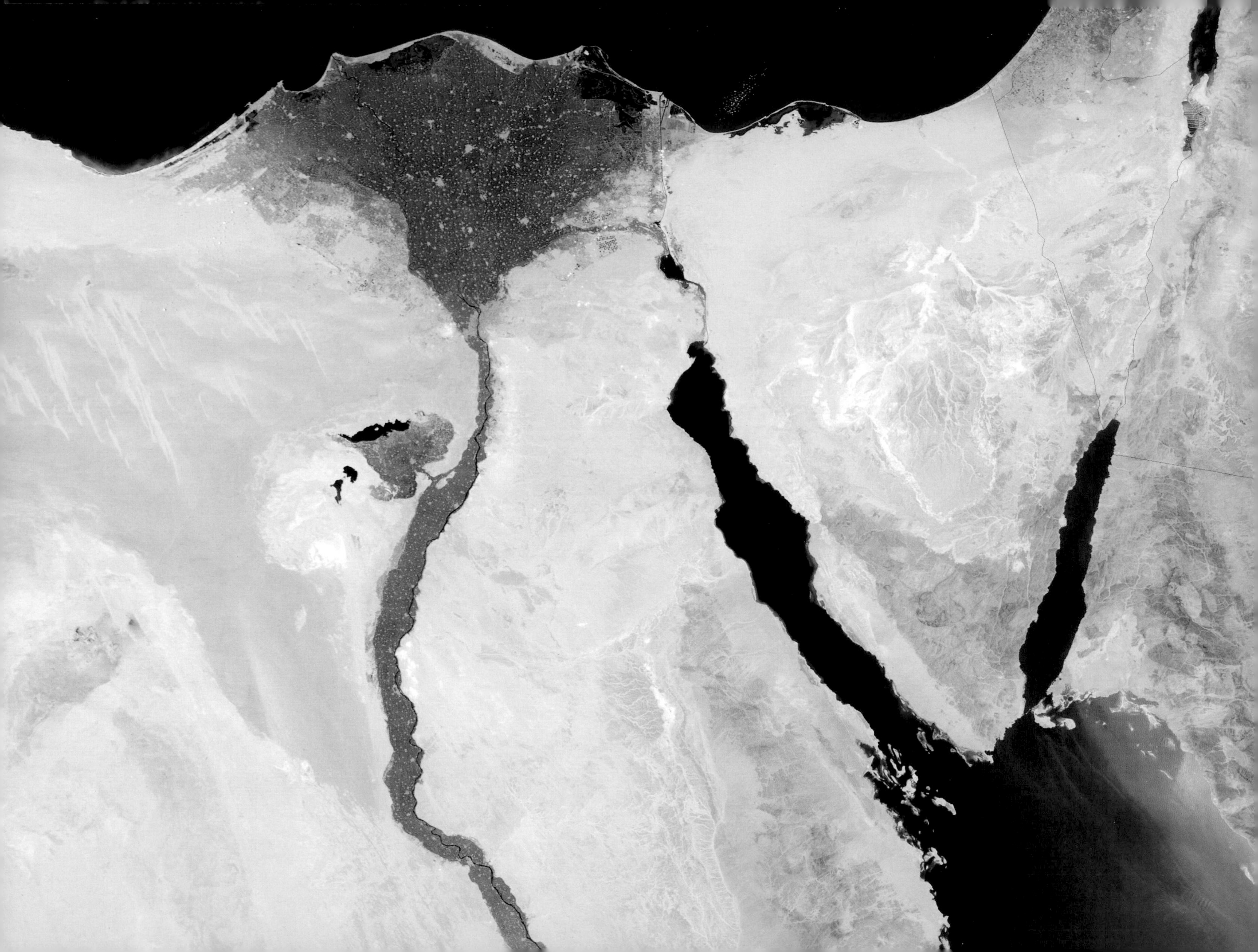

ALL:

**Rivers**

Earth has countless rivers, of which 187 exceed 1000km (600 miles) in length. The longest is the River Nile (pictured opposite), rising from Lake Victoria and discharging into the Mediterranean Sea 6650km (4100 miles) downstream. Most rivers are heavily branched, being joined by tributaries along their length and sometimes dividing into deltas as they reach the sea.

Pictured right is the Delta of Lena River in Siberia, Russia.

**Canyons**

Rivers erode the surface over which they are flowing, which can result in a deep cleft between escarpments or mountains. Such canyons are most common in arid regions where the effects of weathering are localized. Several of Earth's largest canyons are the result of geological uplift of the terrain, into which rivers cut down preserving their meanders in solid rock. Pictured here is the Grand Canyon in Arizona, USA.

## Impact craters

During the early history of the Solar System, planets and their moons were heavily bombarded by asteroids and meteoroids of all sizes. We see evidence of this on the surface of the Moon and can be certain that Earth received a comparable density of impacts. Numerous large impact craters are not evident on Earth today because, unlike the Moon, our planet has an active surface and an atmosphere. Resurfacing caused by plate tectonics together with erosion by weathering has erased most ancient craters from view. Only about 190 impact craters have been identified worldwide, ranging from tens of metres up to 300km (124 miles) in diameter.

Barringer (Meteor) Crater (pictured right) was formed by a meteorite some 50m across which collided with Earth only around 50,000 years ago. The meteorite vaporized on impact and resulted in a crater 1.2km (3/4 mile) wide.

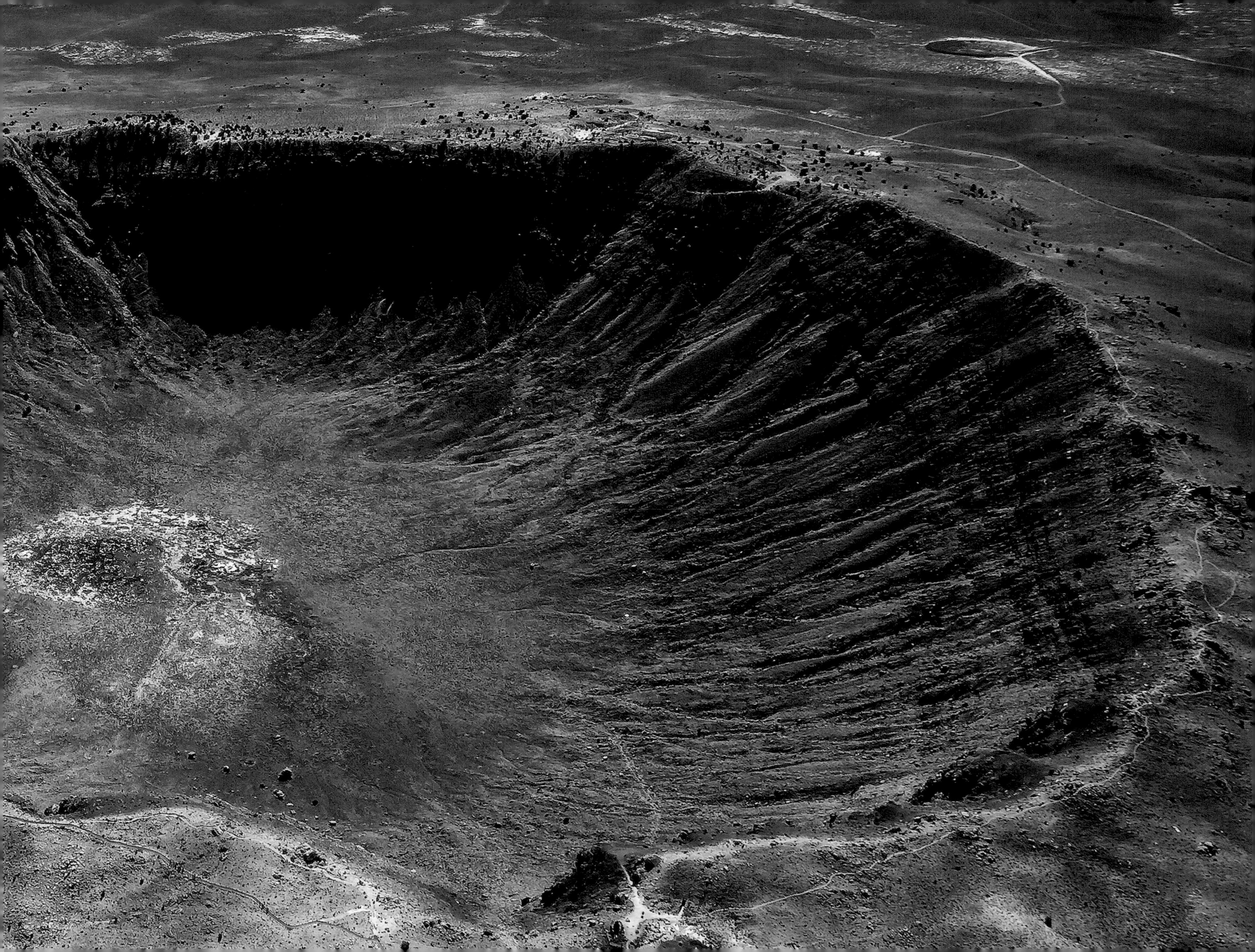

## Magnetic field

In contrast to our planetary neighbours Venus and Mars, a global magnetic field protects Earth from most harmful effects of charged particles emitted by the Sun and cosmic rays from interstellar space. The most visible evidence of Earth's magnetosphere is the aurora, also known as the Northern and Southern Lights. As charged solar particles approach Earth, our magnetic field funnels them towards the north and south magnetic poles. Here they enter the upper reaches of our atmosphere and interact with tenuous gases. As atoms absorb energy they become excited and emit light of characteristic colours.

Earth's magnetic field is generated by a dynamo in our planet's core of molten iron. Without it, life would be exposed to damaging radiation and our planet may not be habitable for complex surface life.

OPPOSITE:

**Atmosphere and climate**

Seen from space, our atmosphere is revealed as a thin blue layer, extending only around 100km (60 miles) from the surface. Three-quarters of the mass of the atmosphere is within just 11km (7 miles) of sea level. Comprising 78 per cent nitrogen and 21 per cent oxygen, it is crucial to maintaining temperatures and pressure at which liquid water can exist on the surface.

Liquid water was present on Earth 3.8 billion years ago, despite the Sun being only 70 per cent as bright then as it is now. It seems that carbon dioxide ($CO_2$) was much more abundant in our early atmosphere, acting as a greenhouse gas to warm the planet. As the Sun has brightened during its lifetime, carbon dioxide levels have decreased, maintaining a fairly stable climate on Earth over billions of years. Much of this carbon was stored in rocks, for example as fossil fuels and subducted into Earth's mantle by plate tectonics. These long-term geological processes led to a pre-industrial level of $CO_2$ of around 280 parts per million. Human activity, particularly burning fossil fuels, has subsequently increased $CO_2$ by 50 per cent to reach 419 parts per million in 2021. This increase in greenhouse gases has resulted in global temperatures increasing by about 1°C in the last century.

RIGHT:

**Weather**

Weather, resulting from the Sun heating the planet's surface, is a fundamental process influencing and shaping Earth. Because the angle of the Sun varies with latitude, different amounts of solar energy are experienced across the globe. This results in the atmosphere having varying temperature, pressure and moisture, which in turn drive winds, cloud, rain, snow and all the other phenomena that constitute weather. In this image we see four tropical cyclones: areas of strong winds spiralling anticlockwise (in the northern hemisphere) around an area of low pressure.

**Biodiversity**

We know from fossils that life began less than one billion years after Earth formed, which suggests that it evolved soon after the physical conditions became suitable. For over two billion years, life remained relatively simple, single-celled organisms.

Life has moulded our planet in many ways. For the first two billion years of Earth's existence, there was little or no oxygen in the atmosphere. The evolution of photosynthesis enabled single-celled organisms called cyanobacteria to utilize the Sun's energy to grow and reproduce, releasing oxygen as a waste product. For billions of years, cyanobacteria have used this energy to build layered structures called stromatolites, as they still do in a few locations today.

Eventually, oxygenation of the atmosphere by photosynthesizing cyanobacteria enabled the evolution of animals, which depend on it for their metabolism. Around 600 million years ago, complex multicellular life became abundant in the fossil record. Since then, at least 99.9 per cent of all species that ever existed have become extinct as a result of natural selection.

Today there are around 1.74 million species of plants, animals, fungi and microorganisms that have been catalogued, with an unknown number of species still to be identified. Around 10 per cent of all current species are thought to live in the Amazon rainforest (pictured left), the most biodiverse habitat on Earth. Forests occupy around 30 per cent of Earth's land area and play a key role in climate regulation, including carbon storage.

**Human population**
Humans are the dominant species on Earth. In terms of total biomass, two other animal species outweigh us (Antarctic krill and domestic cattle) but our effect on the planet is second to none. In one sense, eight billion humans do not take up that much space; most live in urban areas, which occupy less than three per cent of global land area. The photograph shows the city of London from Space.

However, the life support systems for Earth's human population are far more extensive. More than one-third of land area is used for agriculture (the remainder is mostly desert, ice or forest). It is estimated that humans make up 36 per cent of the mass of all mammals on Earth and our livestock represent another 60 per cent. Only four per cent of mammalian biomass is wild species.

ALL:

**Pale blue dot**

The most distant human-made object is the spacecraft *Voyager 1*, which was launched in 1977 (pictured right) and is still operating.

On 14 February 1990, *Voyager* was commanded to turn back towards Earth and take a final picture of its home planet from a distance of 6 billion km (3.7 billion miles, pictured left). Bands of colour in the image are due to sunlight reflected by the camera. Earth is a tiny dot, less than one pixel wide, encompassing everything on our planet and the whole of humanity. At a resolution of one pixel, its predominant colour is blue.

RIGHT:
**Earth and Moon**
The *Deep Space Climate Observatory* orbits the Sun 1.6 million kilometres from Earth to capture images of the day side of our planet. On 16 July 2015, the Moon passed between the satellite and Earth, giving a clear view of their relative sizes.

Earth's diameter is 12,756km (7926 miles), compared to 3745km (2327 miles) for the Moon.

The far side of the Moon looks unfamiliar because we never see it from Earth.

ALL BELOW:

**Near side and far side**

Early on in its history, the Moon's rotation became tidally locked to Earth. The Moon rotates on its axis every 27.3 days, the same time that it takes to orbit Earth. Hence, no matter what its phase, we always see the same face from Earth. The far side was unknown until 1959 when the probe *Luna 3* passed beyond the Moon and transmitted grainy photographs.

The near side of the Moon (pictured below left) is patterned with dark maria, which are basalt plains formed by ancient volcanic eruptions, mostly around three and a half billion years ago. These are largely absent on the far side (pictured below right). The reason for this disparity is still debated but presumably relates to the influence of Earth on the near side causing the lunar mantle to remain molten for longer or its crust to be thinner.

ALL:

**Solar eclipses**

Amongst many curious and intriguing qualities of our Moon is that, as seen from Earth, it appears almost the same size in the sky as the Sun. Whilst the Moon's diameter is just 1/400th that of the Sun, it is on average 390 times closer to us. In fact the Moon's distance varies by about 13 per cent during its orbit, so when nearest (perigee) the Moon is 421 times closer than the Sun and when furthest (apogee) it is 368 times closer. When the Moon passes directly between Earth and Sun, which typically happens twice per year, it eclipses the Sun. If this occurs around lunar perigee the solar disc is completely covered, giving us a total solar eclipse. When the Moon is near apogee a ring of the solar photosphere remains visible, creating an annular solar eclipse.

Pictured above is an annular solar eclipse viewed from United Arab Emirates (26 December 2019), while pictured left is a total solar eclipse viewed from European Southern Observatory, Chile (2 July 2019).

### Lunar craters

The Moon has hundreds of thousands of craters larger than 1km (0.6 miles) in size and innumerable smaller ones. Nearly all were formed by asteroid impacts rather than volcanoes. Many originated billions of years ago when impacts were much more frequent than today. They remain well-preserved owing to the lack of atmospheric weathering or resurfacing by subsequent geological processes.

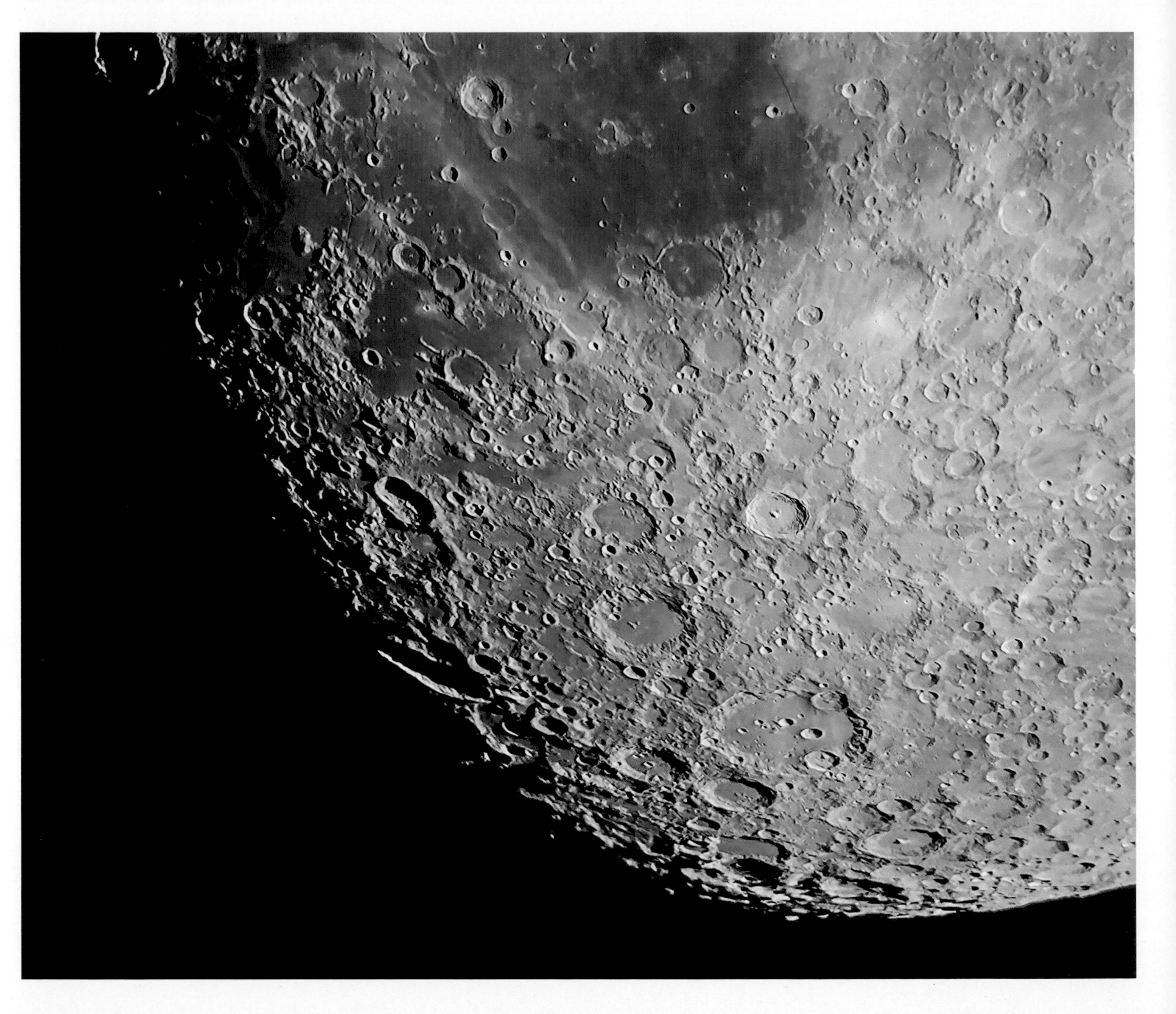

RIGHT: The deeply cratered lunar southern highlands, with the prominent crater Tycho at the centre of the image. Tycho is relatively young so its edges are sharply defined, with few subsequent impacts to degrade them.

OPPOSITE: Tycho is surrounded by bright 'rays' formed by material thrown out by the original impact. These were sampled by the crew of *Apollo 17*, giving an estimated age for the crater of 108 million years.

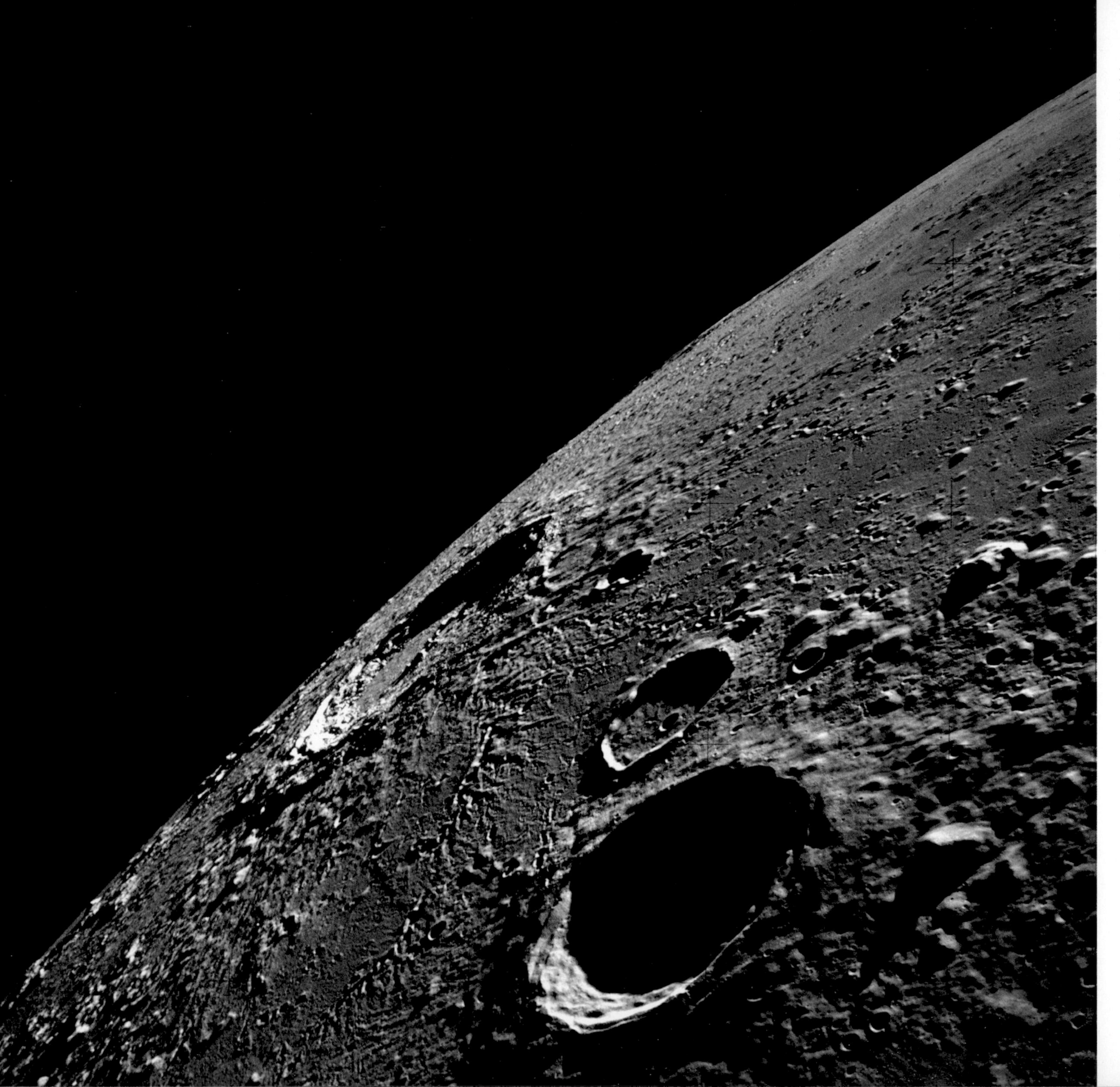

**Lunar craters**

LEFT: Daedalus, a major crater near the centre of the lunar far side is never visible from Earth. It was photographed from lunar orbit by the crew of *Apollo 11*.

OPPOSITE: Clavius is the second-largest crater on the near side of the Moon. It was formed billions of years ago and is overlain by numerous subsequent impact craters.

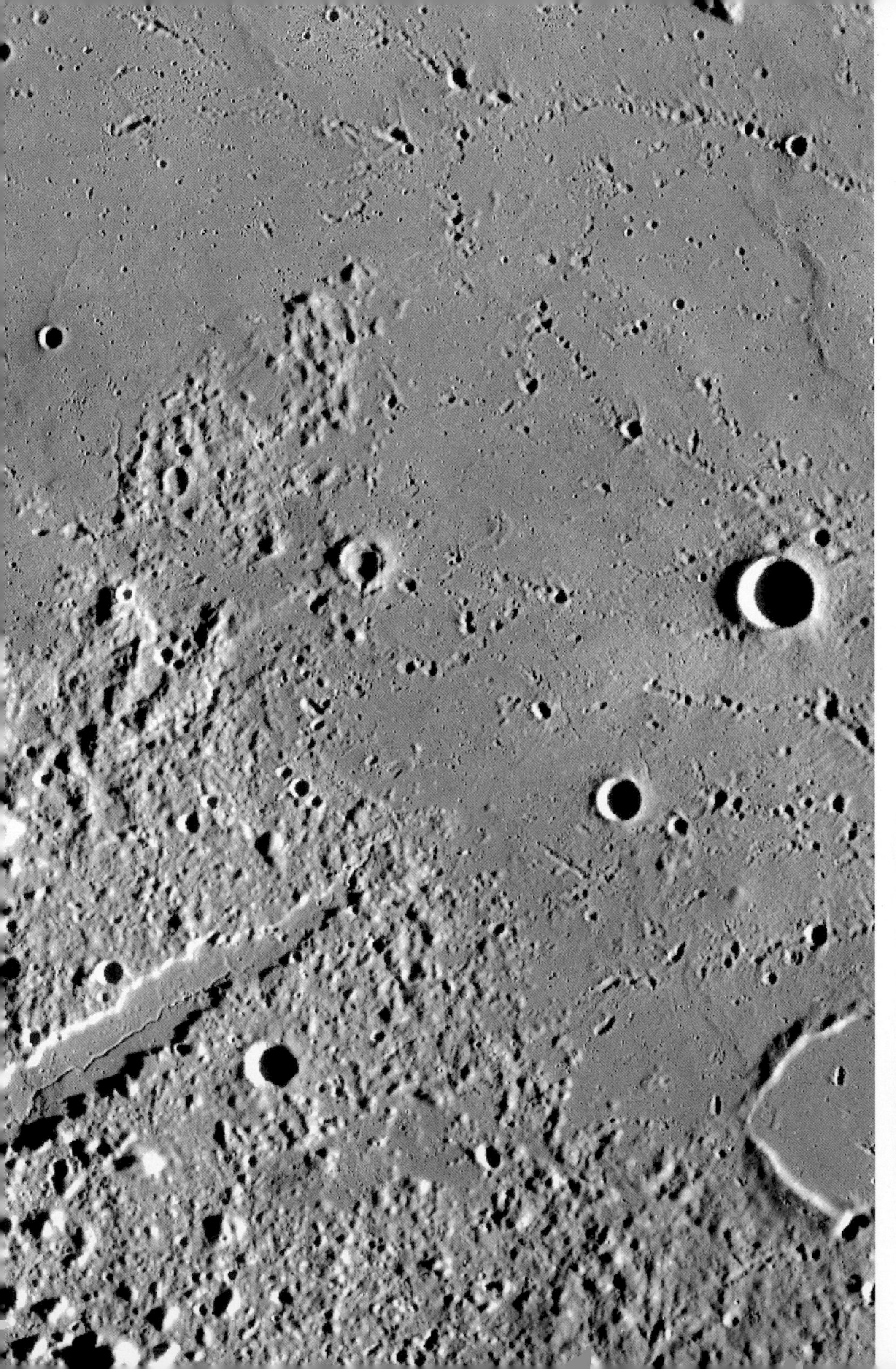

**Lunar mountains**

The Moon has no tectonic plates or active volcanoes. Its mountains result from asteroid impacts ejecting material into piles of rubble. Many are named after mountain ranges on Earth but usually in Latinized form. In the absence of a sea level, lunar mountains are measured from a baseline defined relative to the centre of the Moon.

LEFT: Montes Alpes stretches about 280km (174 miles), with peaks ranging from 1800m to 2400m (3700–7800ft) in height.

BELOW: Montes Apenninus includes Mons Huygens, around 5500m in elevation, which is usually described as the highest lunar mountain (though not the highest point on its surface).

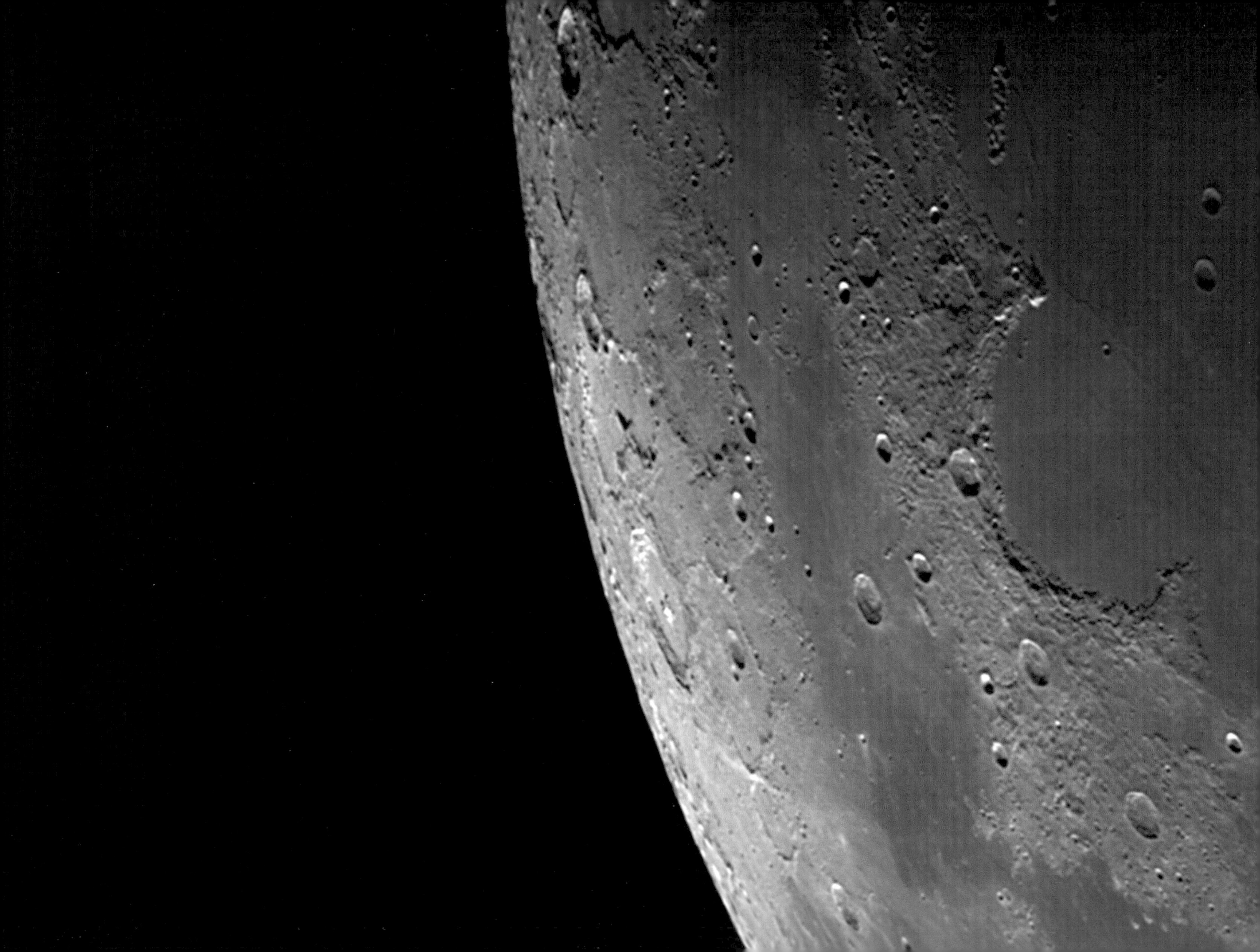

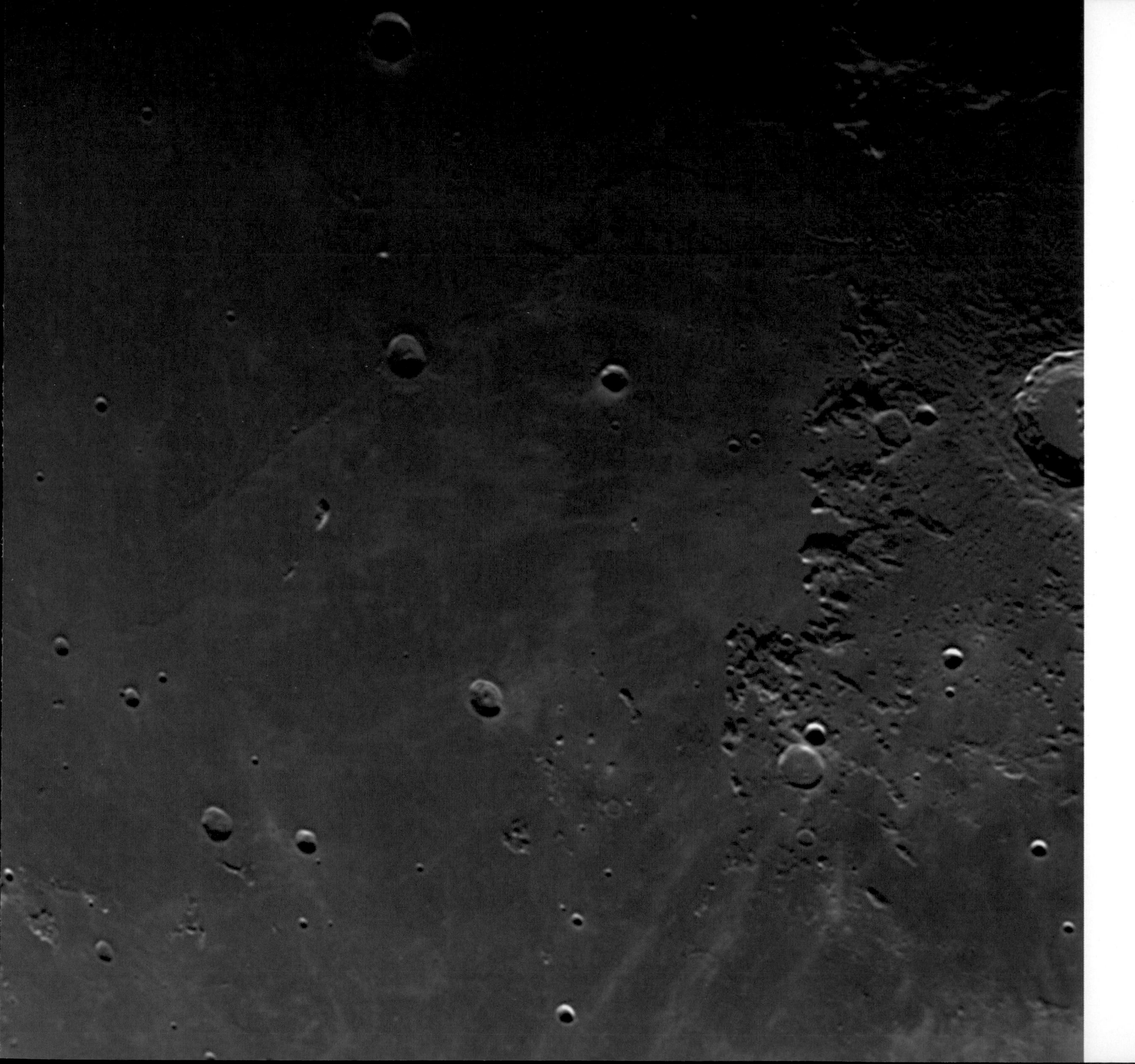

**Lunar maria**

Lunar maria (singular mare) are dark plains formed when lava was extruded from the lunar mantle and solidified into basalt. Most of them occupy basins created by asteroid impacts.

Their name 'maria' is Latin for 'seas' and was given because early observers incorrectly assumed them to be filled with water. Smaller maria are referred to as 'sinus' meaning 'bays'. There is no volcanic activity on the Moon today, though there may still be some magma deep beneath the surface.

Pictured left is Mare Imbrium, with semicircular Sinus Iridum occupying a large impact crater in the centre of the image. The mountainous ridge of Montes Jura forms an arc around the right-hand side of Sinus Iridum.

ALL:
**Lunar air, soil and water**
Being smaller than Earth, the Moon's surface gravity is just one-sixth as much. That means that a person weighing 62kg (137lbs) on Earth would weigh just over 10kg (22lbs) on the Moon. Lunar gravity is too small to retain any significant atmosphere, so the Moon is to all intents and purposes an airless, barren world. However, rock samples from the *Apollo* missions show that there was once a tenuous atmosphere, as a result of volcanic outgassing around three and a half billion years ago. Those gases have long-since escaped to space, stripped away by the solar wind as the Moon lacks a protective magnetic field. In the absence of an atmosphere, weathering processes on the Moon are limited to the solar wind and cosmic rays.

Temperatures range from 140°C (284°F) on the day side to -171°C (-275°F) at night. The lunar surface is shaped by impacts of large and small meteoroids, which have ground the surface into particles and dust called regolith. Liquid water cannot persist on the lunar surface.

However, permanently shadowed craters near the north and south poles have been found to contain water ice, in much the same way as on Mercury.

**The origin of the Moon**
Lunar rocks collected by *Apollo* astronauts were found to have almost identical isotopic compositions to those of rocks on Earth, indicating that both formed from the same material. Radioisotope dating indicates that the Moon formed about 50 million years after the Solar System. These observations, and the high angular momentum of the Earth–Moon system, cannot be accounted for if both had formed simultaneously from the same interplanetary disc or if the Moon had formed independently and then been captured by Earth's gravity.

The accepted explanation is that the primordial Earth was involved in a collision with a smaller planet (named Theia), which disintegrated as a result (shown in this artist's impression right). Material thrown out from the impact coalesced to form the Moon, initially in an orbit around 10 times closer than it is today. Tidal interactions over 4.5 billion years have caused its orbit to expand to its current average distance of 383,000km (238,000 miles). Those same forces have reduced the rate of Earth's rotation from an estimated five hours when the Moon was formed to our current day length of 24 hours. These processes continue today; the Moon is receding from Earth by 3.8cm (1.5in) annually and the Earth's rotation is slowing by 23 microseconds per year.

LEFT, OPPOSITE & OVERLEAF:
**Exploring the Moon**
On 21 July 1969, Neil Armstrong and Edwin "Buzz" Aldrin became the first humans to set foot on the Moon. They were followed by 10 more astronauts in the following three-and-a-half years. In the 50 years since, no human has travelled beyond low Earth orbit. The successful Moon landing and return to Earth of the crew of *Apollo 11* remains one of the greatest technological, cultural and human exploration feats ever accomplished.

Armstrong and Aldrin spent just two hours on the lunar surface, during which time they recorded extensive observations, set up scientific experiments, took many photographs and gathered 22 kilograms of rock and dust samples. Analysis of the lunar material they brought back to Earth provided evidence in support of the giant impact theory of Moon formation.

LEFT:
**Earthrise**
*Apollo 8* was the first crewed spaceflight to travel far enough to view the Earth in its entirety. On Christmas Eve 1968, the spacecraft was in orbit around the Moon and astronaut Bill Anders saw the chance to photograph the partly illuminated Earth coming into view over the lunar horizon. The contrast between the grey, lifeless Moon and our blue planet with its oceans and atmosphere teeming with life had an instant impact around the world.

# Mars

Noticeably red in hue even to the unaided eye, Mars was named by the Romans after their god of war. Featuring in both science fiction and popular culture, the fourth planet from the Sun has long been a source of fascination. Since 1964, Mars has been visited by around 48 spacecraft launched by seven countries, 11 of which have successfully landed on the planet.

Mars has some notable similarities to Earth. Its rotation period is 24 hours 37 minutes and its axial tilt is 25°, so day length and seasons are comparable. In many respects, however, Mars is a dramatically different world from our own. It is about half the diameter and one-ninth as massive, so its surface gravity is much weaker at just 38 per cent that of Earth.

Mars has been unable to retain a thick atmosphere, with the result that surface atmospheric pressure is less than one per cent as dense. This means that liquid water cannot persist on the surface. However, there is abundant evidence that conditions were different in the distant past.

Like the Moon and Mercury, the surface of Mars is heavily cratered by meteorite and asteroid impacts, indicating that much of it is billions of years old. Volcanic features cover large parts of the planet, some of them originating within the last 20 to 200 million years. Marsquakes have been recorded, showing that it is still geologically active. Rocks exhibit evidence of past tectonic activity and a former global magnetic field. However, the internal dynamo generated by Mars's metallic core, which is about half the size of Earth's, ceased billions of years ago.

Today the planet has no plate tectonics and no internal magnetic field. Despite several searches by robotic probes, no unambiguous signs of Martian life have been found. A key question for current research is whether such life existed in the past.

OPPOSITE:

**Mapping Mars**

*Mariner 9* became the first spacecraft to orbit Mars in 1971, followed by two *Viking* orbiters in 1976. Between them they imaged most of the planet. This view of the planet's western hemisphere shows giant volcanoes, what appears to be a huge canyon and evidence of former rivers on the surface.

RIGHT:

**The first close-up view of Mars (1965)**

In July 1965, after a journey of nearly eight months, *Mariner 4* became the first spacecraft to fly past Mars. A total of 22 images were transmitted to Earth with maximum resolutions of about 1km, revealing a cratered, apparently desolate world. The Martian atmosphere was found to be much thinner than expected. Mars seemed much more similar to the Moon than to Earth, ending hopes of finding intelligent life on the Red Planet.

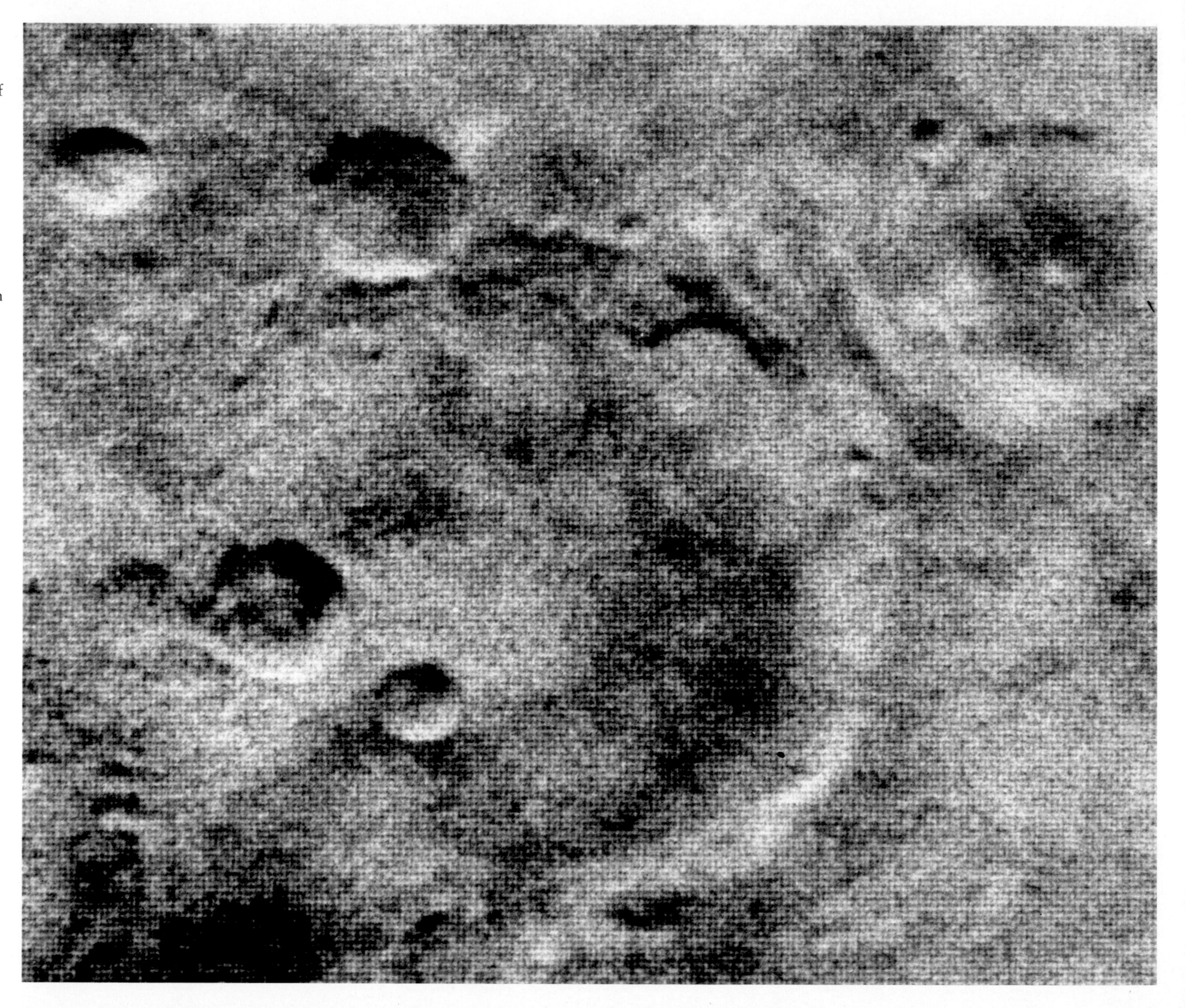

LEFT:

**Mars photographed by Hubble Space Telescope**

A Martian year is 687 Earth days long, with the result that the two planets come close together every 26 months. At these oppositions, detail can be seen on the surface of Mars using Earth-based telescopes. This image was captured by *Hubble Space Telescope* in Earth orbit during the opposition of 2001, at a distance of 67 million km (42 million miles). The planet's distinctive red colour is because its surface rocks are rich in iron oxide.

LEFT:

**The first landing on Mars**
On 20 July 1976, *Viking 1* touched down at Chryse Planitia, becoming the first spacecraft to successfully land on Mars. It was followed six weeks later by *Viking 2* at Utopia Planitia. These pictures were amongst the first returned from the surface of Mars. The primary purpose of both landers was to search for evidence of Martian life using three biological experiments and a spectrometer to measure the presence of any organic compounds in the soil.
The results, however, were inconclusive. Only one of the four experiments gave a positive result, which many scientists think may have been due to inorganic chemical reactions.

RIGHT:

**Martian climate**

Compared to Earth, Mars averages 52 per cent further from the Sun, so it receives less than half the amount of sunlight and a Martian year takes twice as long to complete. However, the Red Planet's orbit is much more elliptical, so its distance from the Sun ranges from 207 million km (during its southern hemisphere summer) to 249 million km (during its northern hemisphere summer). Surface temperatures range from a chilly -143°C (-225°F) at the poles in winter to highs that can reach 35°C (95°F) near the equator in the southern hemisphere summer.

Like Earth, the Red Planet has permanent polar ice caps. Our understanding of these has been greatly enhanced by *Mars Express* (in orbit since 2003) and *Mars Reconnaissance Orbiter* (since 2006), both of which pass directly over the polar regions. This image (right) shows the north polar ice cap.

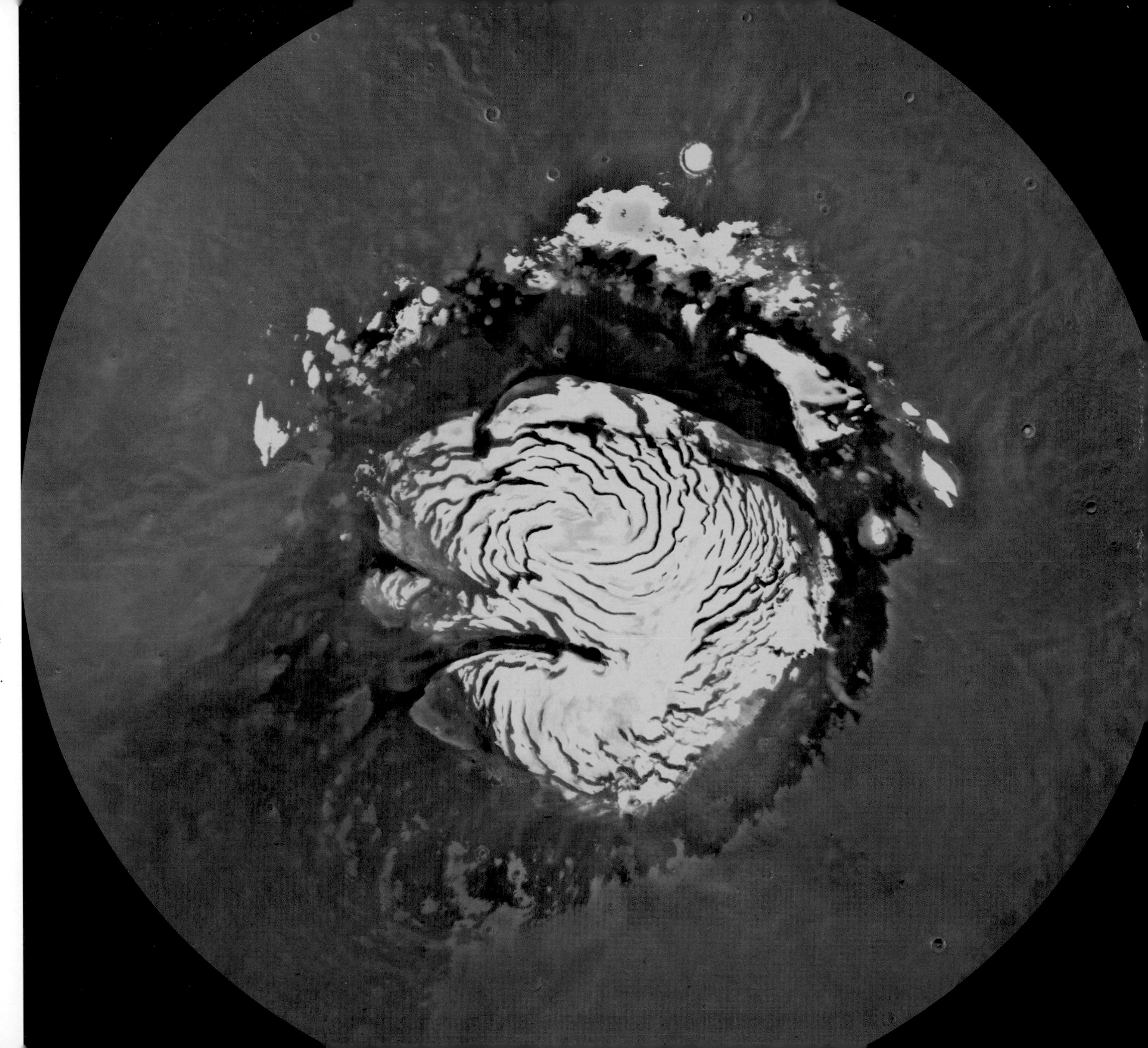

ALL:

**Martian polar ice caps**
The ice caps of Mars largely comprise frozen water. They differ from Earth's in that during the winter when the poles are in darkness, carbon dioxide freezes from the Martian atmosphere to form a layer of dry ice. In spring and summer, this carbon dioxide sublimates directly from ice to gas. The ice caps are shown in these two remarkable perspective images captured by *Mars Express*.

The northern polar ice cap (above right) is around 1000km (600 miles) in diameter and 2km (1.2 miles) thick, whilst the southern ice cap (below right) is 350km (200 miles) in extent and 3km (2 miles) in depth. Between them, the ice caps of Mars contain more frozen water than the Greenland ice cap on Earth. The *Mars Reconnaissance Orbiter* image opposite shows detail in south pole ice sheets.

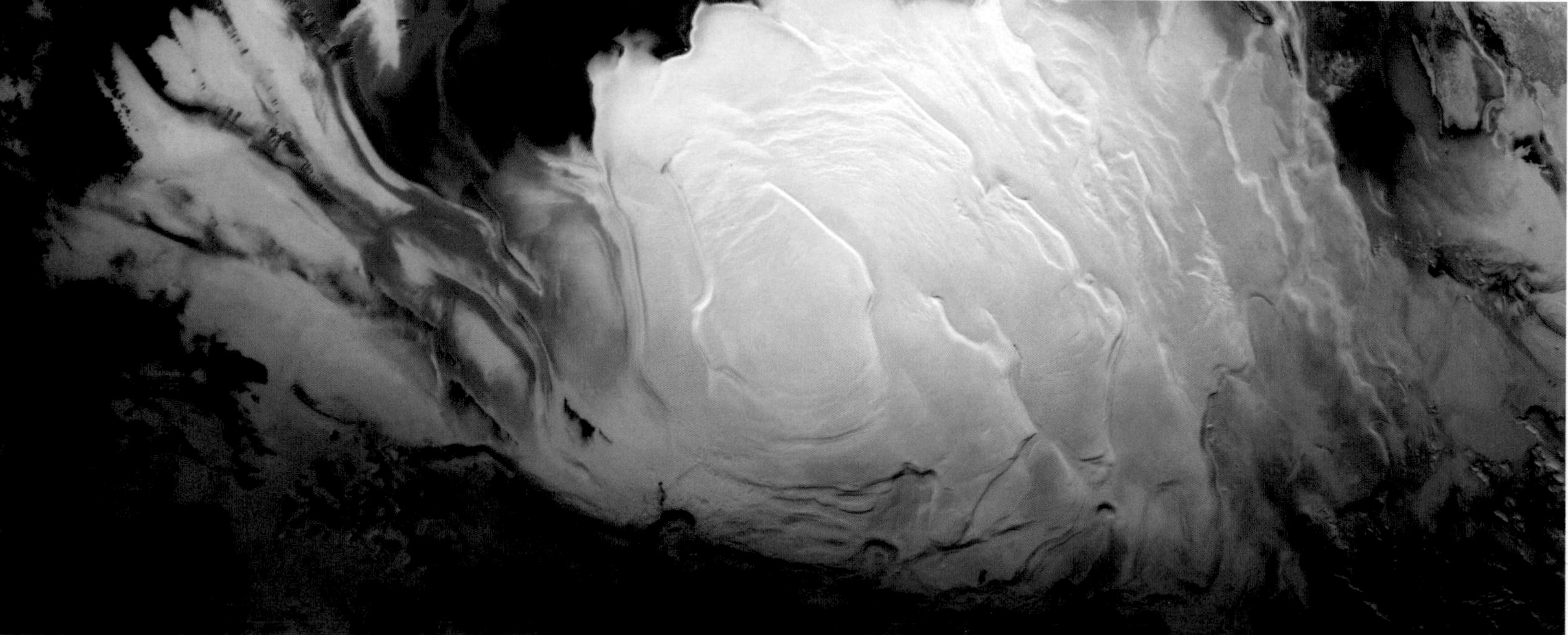

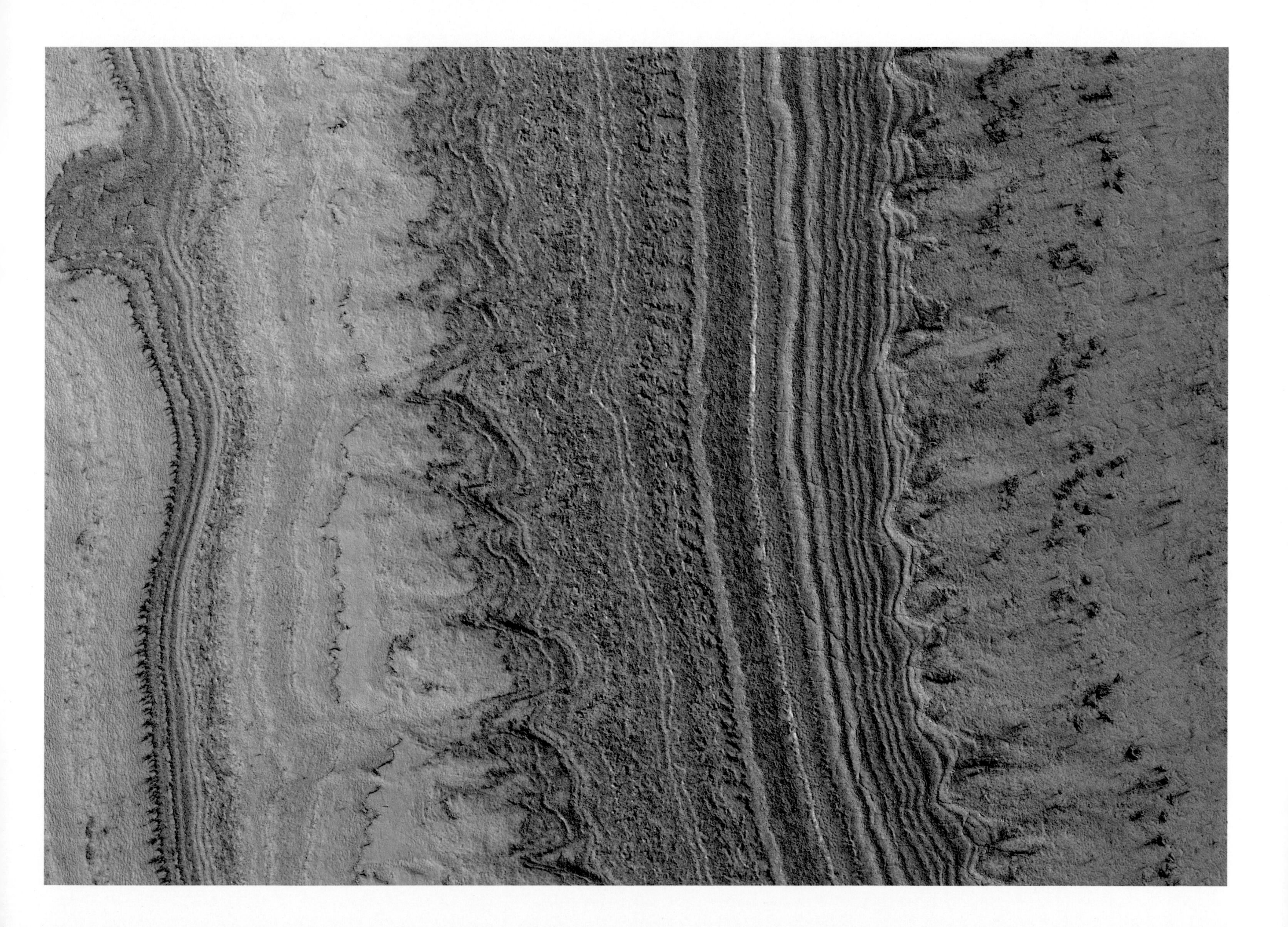

RIGHT:

**Spring in the southern polar region of Mars**
Araneiform ('spider-like') features formed by dry ice as it sublimates in spring in the south polar region (*Mars Reconnaissance Orbiter*).

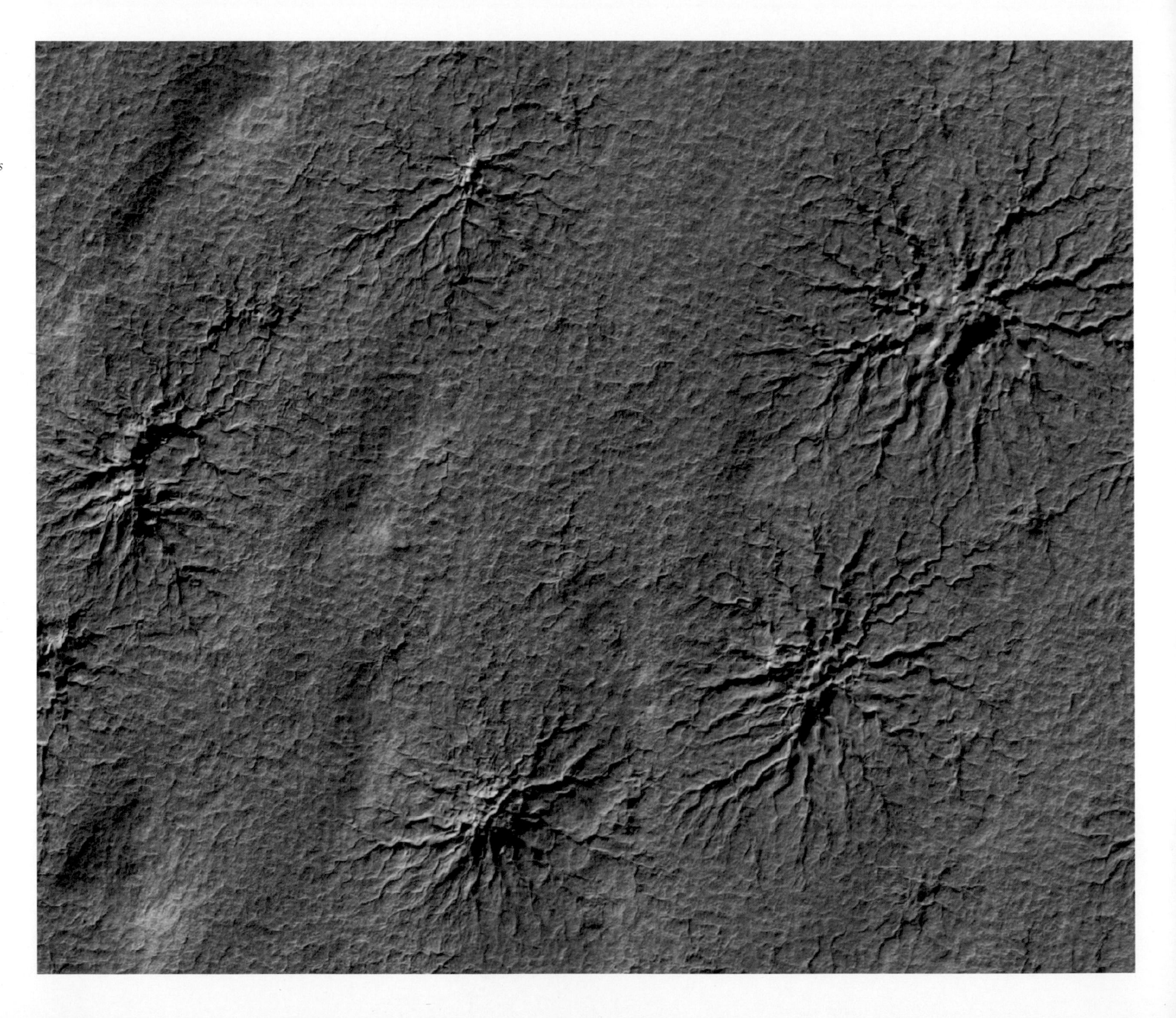

LEFT:

**Martian craters**

There are estimated to be more than 43,000 Martian impact craters over 5km (3 miles) in diameter, and perhaps a quarter of a million comparable in size to Barringer (Meteor) Crater on Earth. This is direct evidence that the surface of Mars is much older than that of Earth and Venus, with less tectonic activity and atmospheric weathering.

The 800m (1/2 mile) wide Victoria Crater, shown here in a stunningly detailed image from *Mars Reconnaissance Orbiter*, has a delicate pattern of sand dunes on its floor.

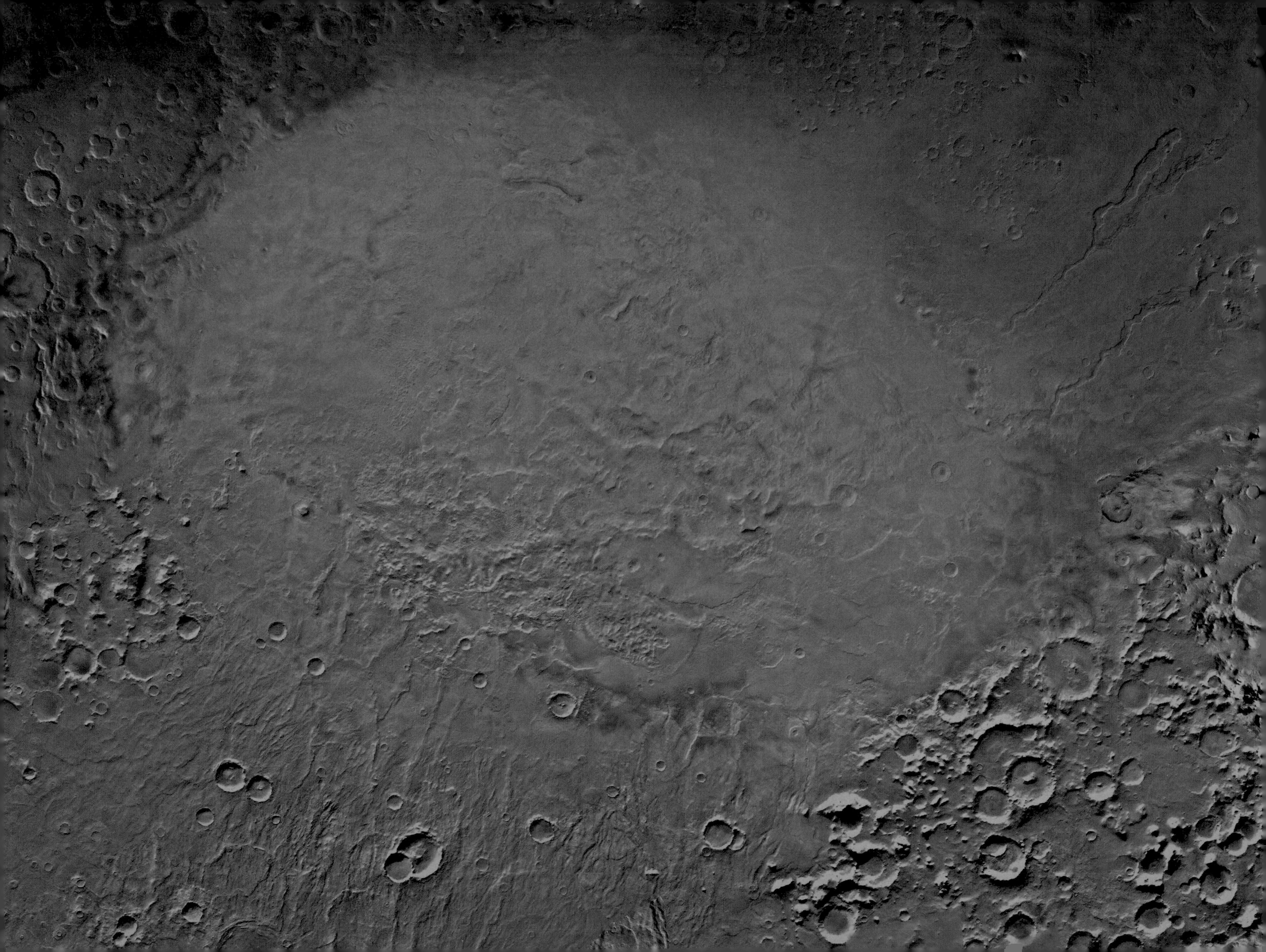

LEFT:

**Hellas Planitia**

Hellas Planitia is the biggest unambiguous Martian impact crater and one of the largest in the Solar System. Thought to have been formed around four billion years ago, it is a huge depression some 2300km (1400 miles) in diameter and 7km (four miles) deep; making it the lowest point on the planet's surface.

BELOW:

**Korolev Crater**

Many craters, including Hellas Planitia, show evidence of past glaciation. In the polar regions, these processes can be observed continuing today. Korolev, at latitude 73°N, is 81km (50 miles) in diameter and contains over 2000km$^3$ of water ice.

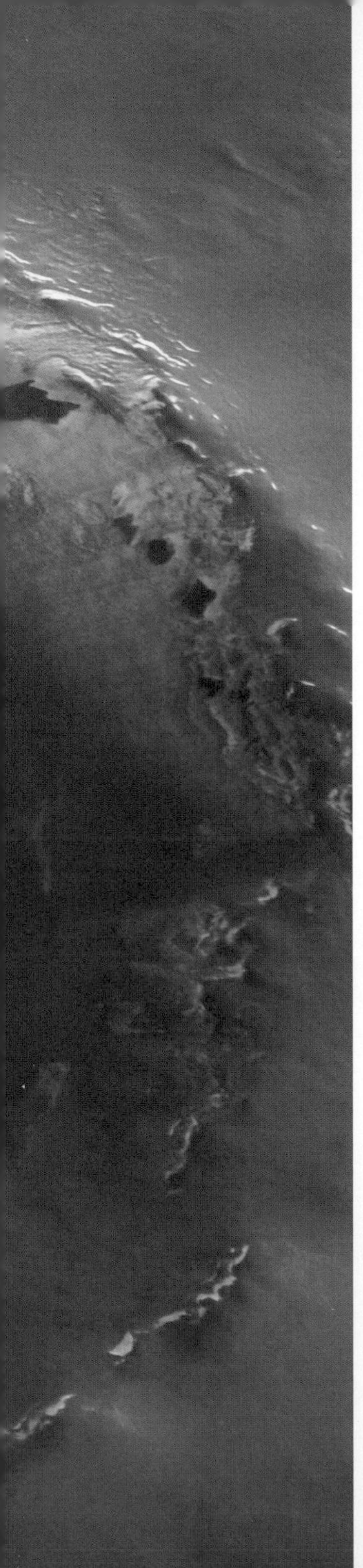

LEFT:
**Crater ice**
*Mars Express* obtained this view of an unnamed impact crater located on Vastitas Borealis in Mars's far northern latitudes. The circular patch of bright material located at the centre of the crater is residual water ice. This patch of ice is present all year round, remaining after frozen carbon dioxide overlaying it disappears during the Martian northern hemisphere summer.

RIGHT:
**New craters**
Craters are continually being formed on Mars. *Mars Reconnaissance Orbiter* photographed this one in February 2019; it was not present in a previous image taken in September 2016.

## Volcanoes

Past volcanic activity has shaped much of the Martian surface but no active volcanoes have been observed. Like Earth, Mars is thought to have a molten iron-rich core but being a smaller planet it has cooled more rapidly, so convection in the mantle shut down billions of years ago. It seems that magma no longer reaches the surface, or does so rarely.

The highly elevated Tharsis region in the planet's western hemisphere has five huge extinct volcanoes, each much bigger than any found on Earth. The lower gravity of Mars results in fewer but larger eruptions of lava. Moreover, the lack of plate tectonics results in the Martian crust being stationary over plumes of magma in the mantle, unlike on Earth where the crust is in constant motion. These two factors have led to the formation of colossal volcanoes.

With a diameter of 600km (370 miles) and a height of 21km (13 miles), Olympus Mons is the largest volcano in the Solar System (pictured right). It is a shield volcano, similar in structure to Mauna Loa in Hawaii but twice the height and almost 100 times bigger in volume.

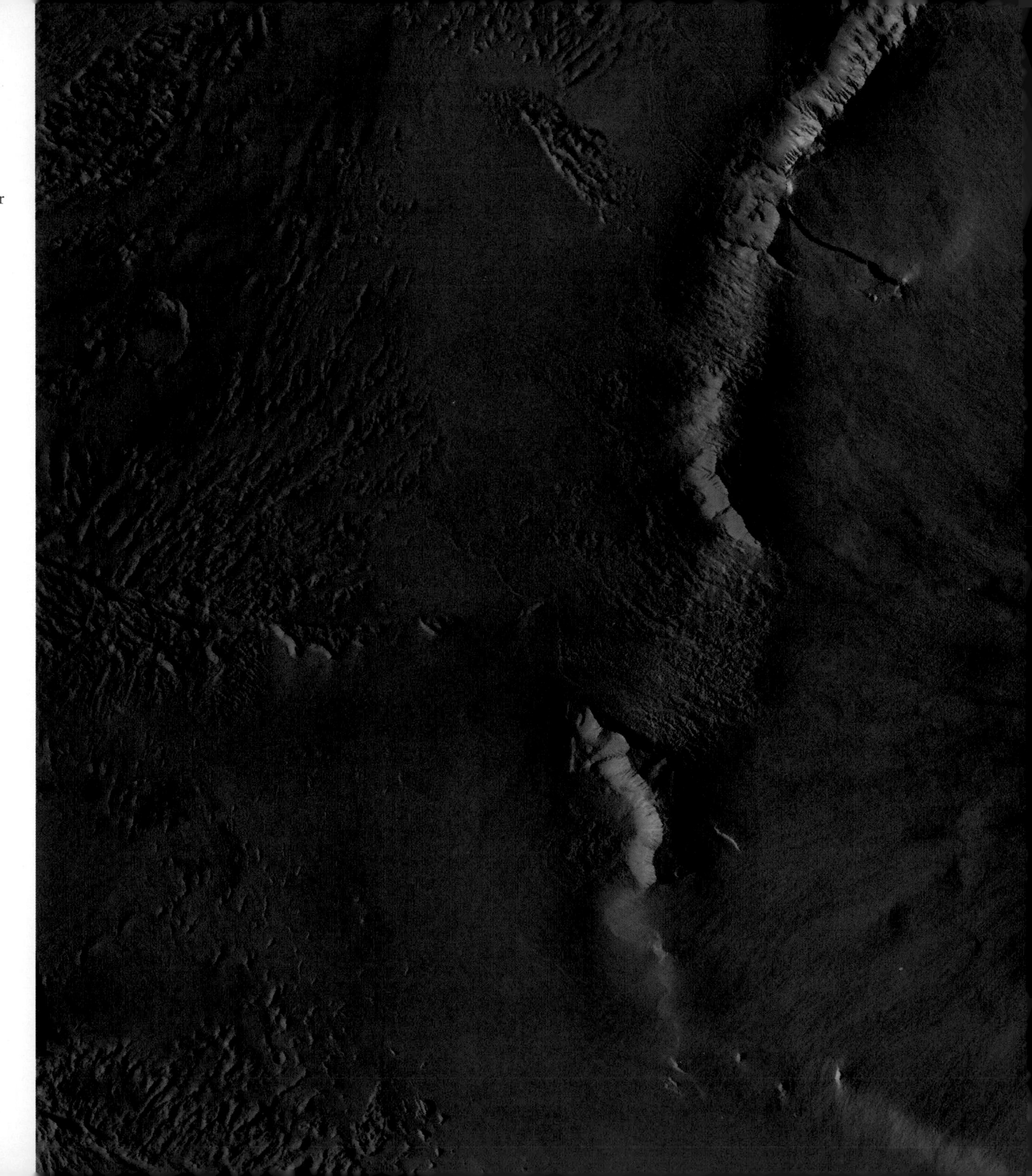

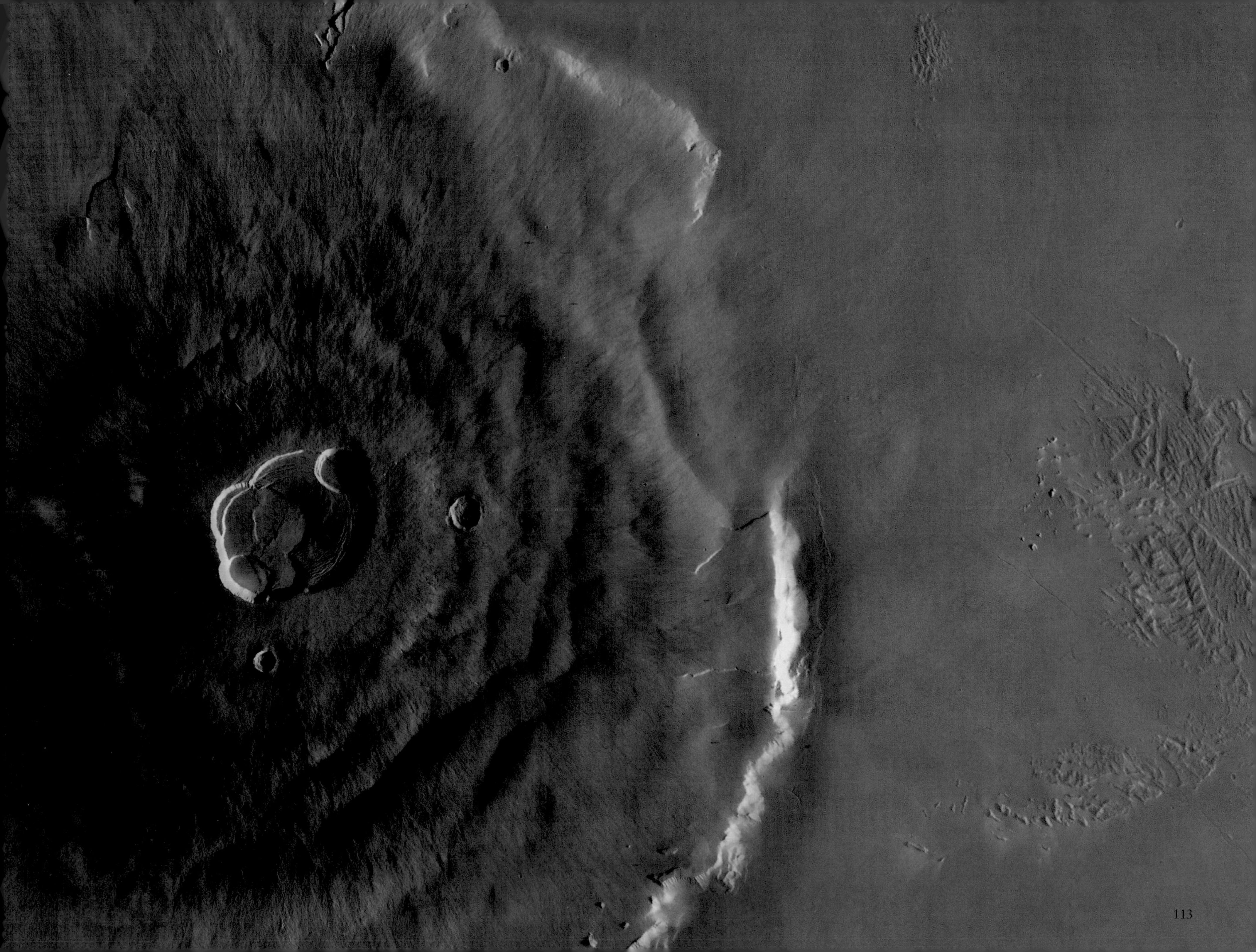

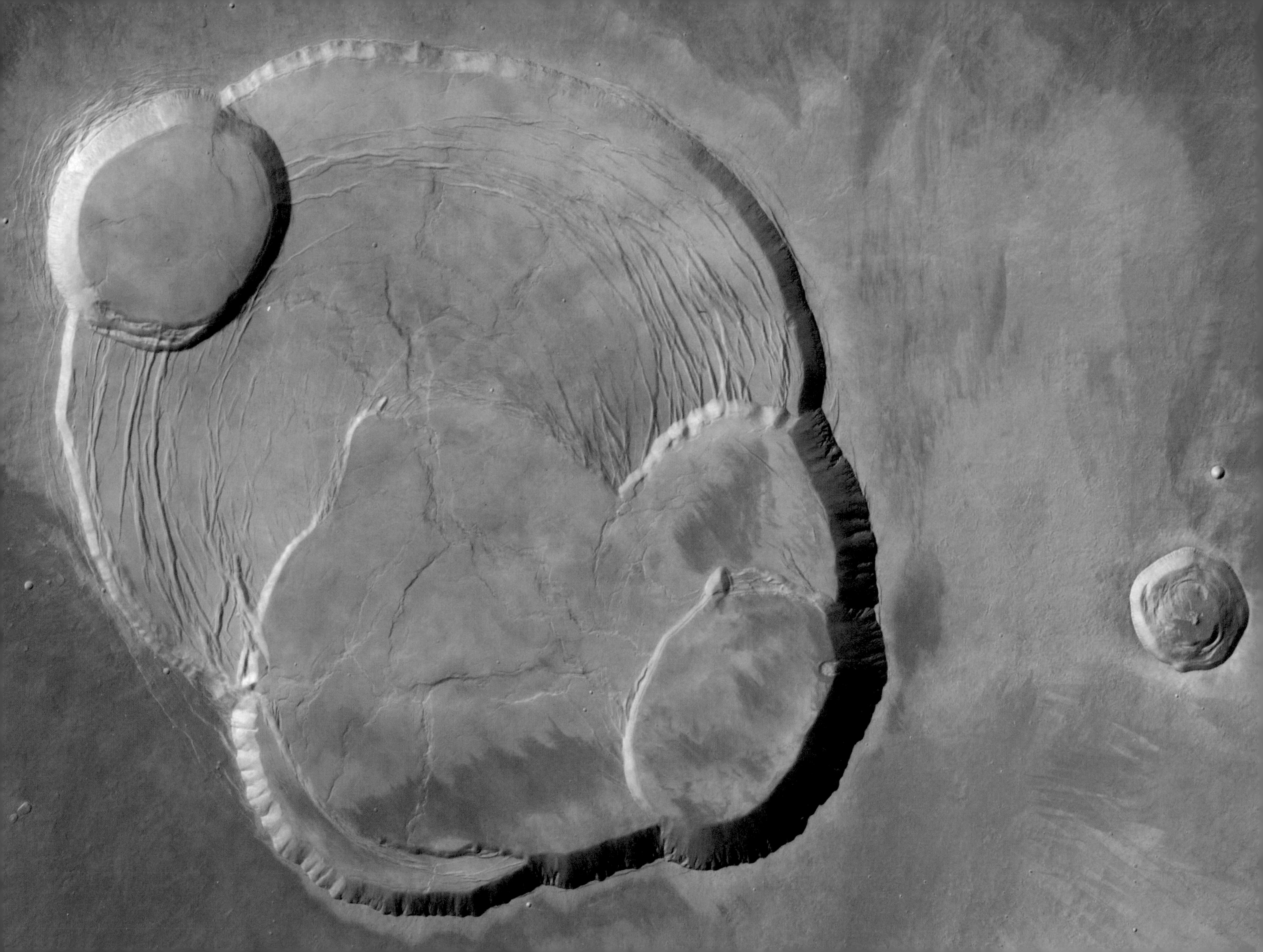

ALL:
**Olympus Mons**
Olympus Mons was first seen by telescopes on Earth, from where nineteenth century astronomers deduced it to be a high mountain as the summit could be discerned even when the rest of the Martian surface was obscured by dust storms. Crater counts suggest that the most recent eruptions of Olympus Mons took place within the last 100 million years, which is very recent in the geological history of Mars.

These close-up images of the calderas at the summit of Olympus Mons were obtained by *Mars Express*.

ALL:
**Ancient volcanic activity**
Extinct volcanoes and ancient eruptions have sculpted much of the Martian surface. The image on the left shows a possible cinder cone on the huge shield volcano of Pavonis Mons. On the opposite page is an old lava flow in the Tempe Terra region.

ALL:

**Atmosphere and weather**

Atmospheric pressure on the surface of Mars is less than one per cent of that on Earth, equivalent in density to air at an altitude of 35km (21 miles) above terrestrial sea level. Moreover, the Martian atmosphere is almost entirely carbon dioxide, so it is toxic to terrestrial animal life. Although carbon dioxide is a greenhouse gas, the air is so thin that it retains little heat. Average surface temperature on Mars is -58°C (-72°F), comparable to that of inland Antarctica.

Despite being so tenuous, the Martian atmosphere creates significant weather. High altitude ice clouds form over the Tharsis volcanoes, including Olympus Mons, as a result of air being forced upwards by the mountains and cooling. Ice crystals can precipitate as snow. Localized dust storms are a frequent occurrence and have been observed to become global in extent every few years.

Pictured left are orographic ice clouds above the Tharsis volcanoes. The image on the opposite page shows a dust devil, estimated to be 70m (230ft) wide and 20km (12 miles) high, on Amazonis Planitia.

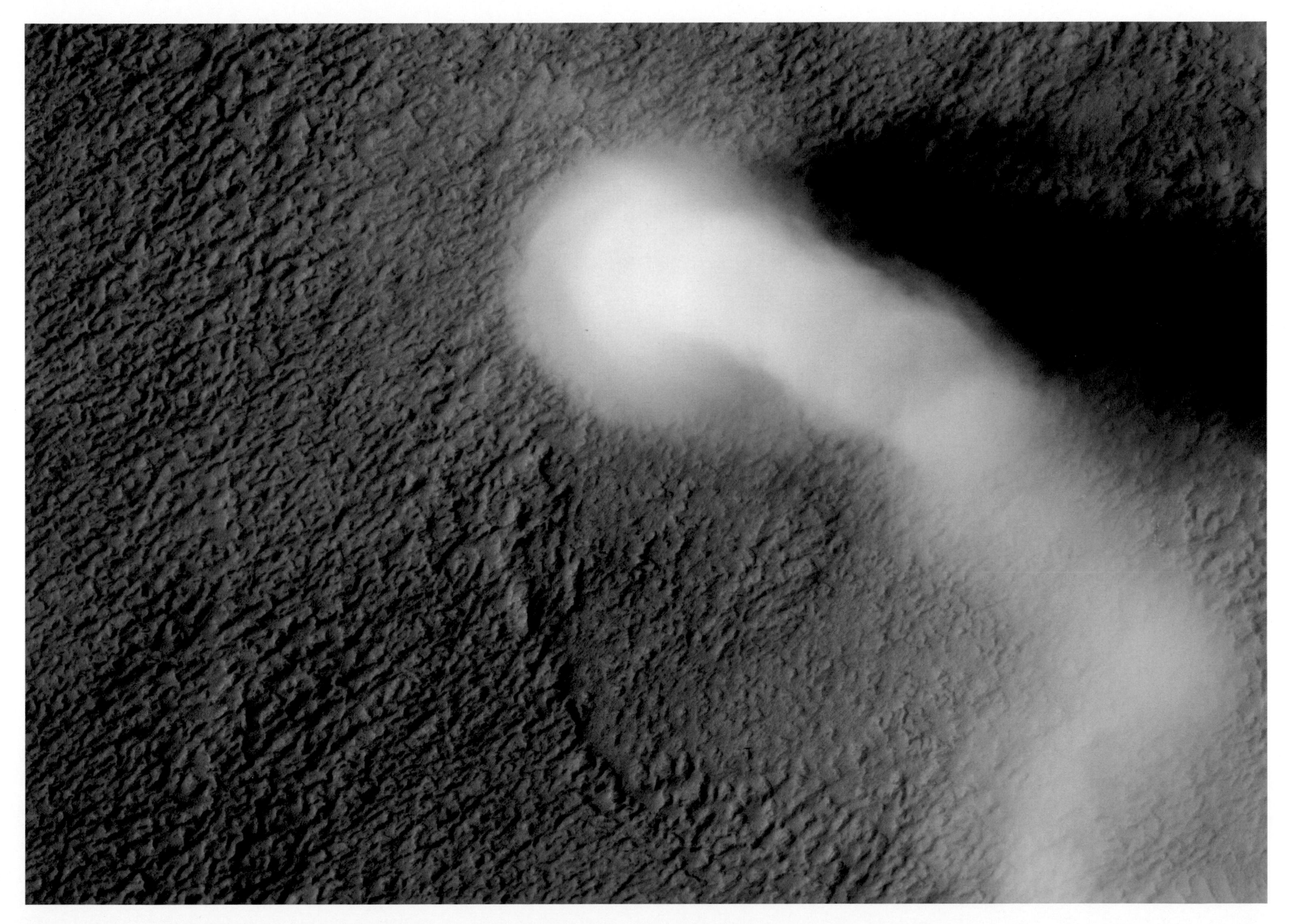

LEFT:
**Dust storms**
Mars periodically experiences planet-wide dust storms. When *Mariner 9* arrived in November 1971, the surface was totally obscured by dust and it was several weeks before it could commence its imaging programme. This image from *Mars Global Surveyor* shows localised dust storms on Syria Planum.

RIGHT:
**Sand dunes**
Mars is a desert world. However, it may seem surprising that the thin Martian atmosphere creates wind-formed landscapes. This dune field near the north polar ice cap was photographed by *Mars Reconnaissance Orbiter*.

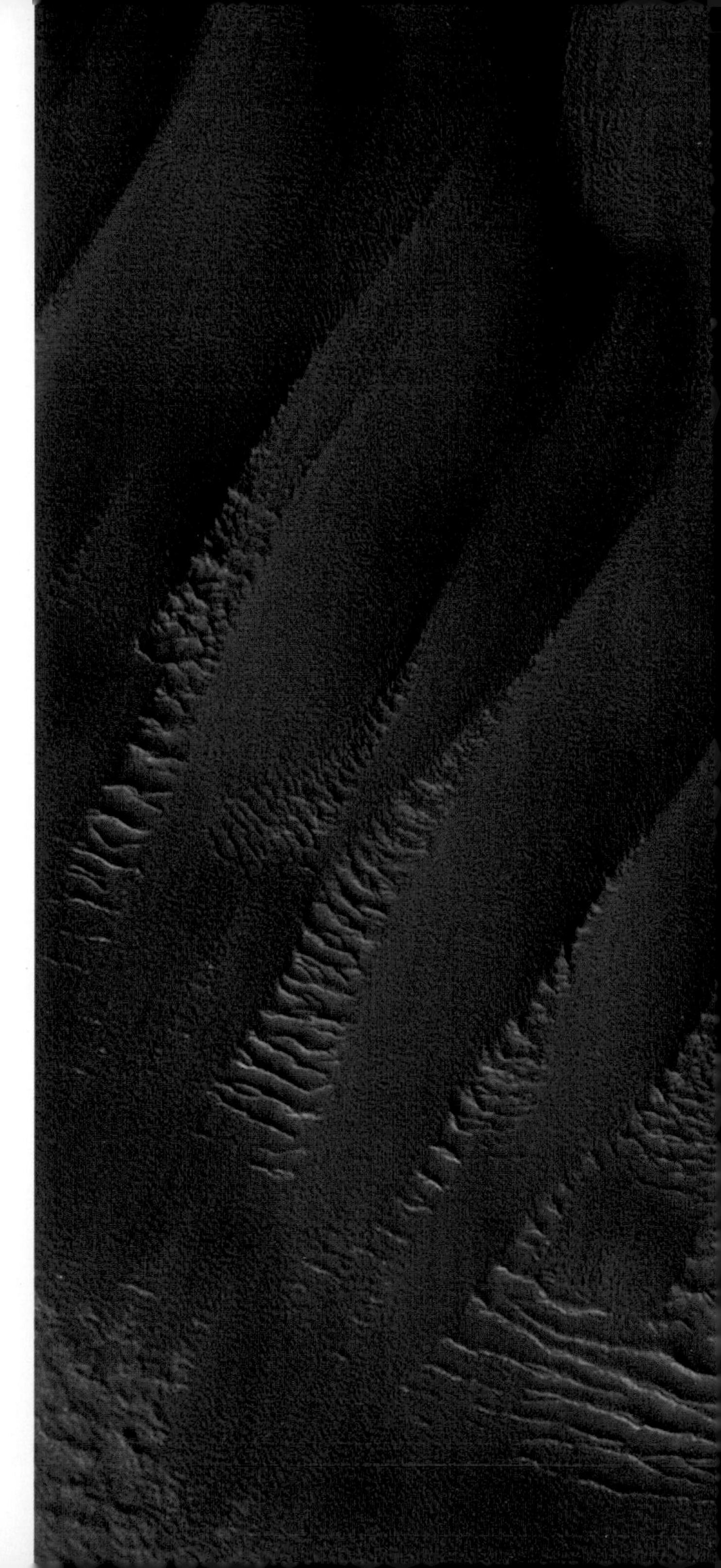

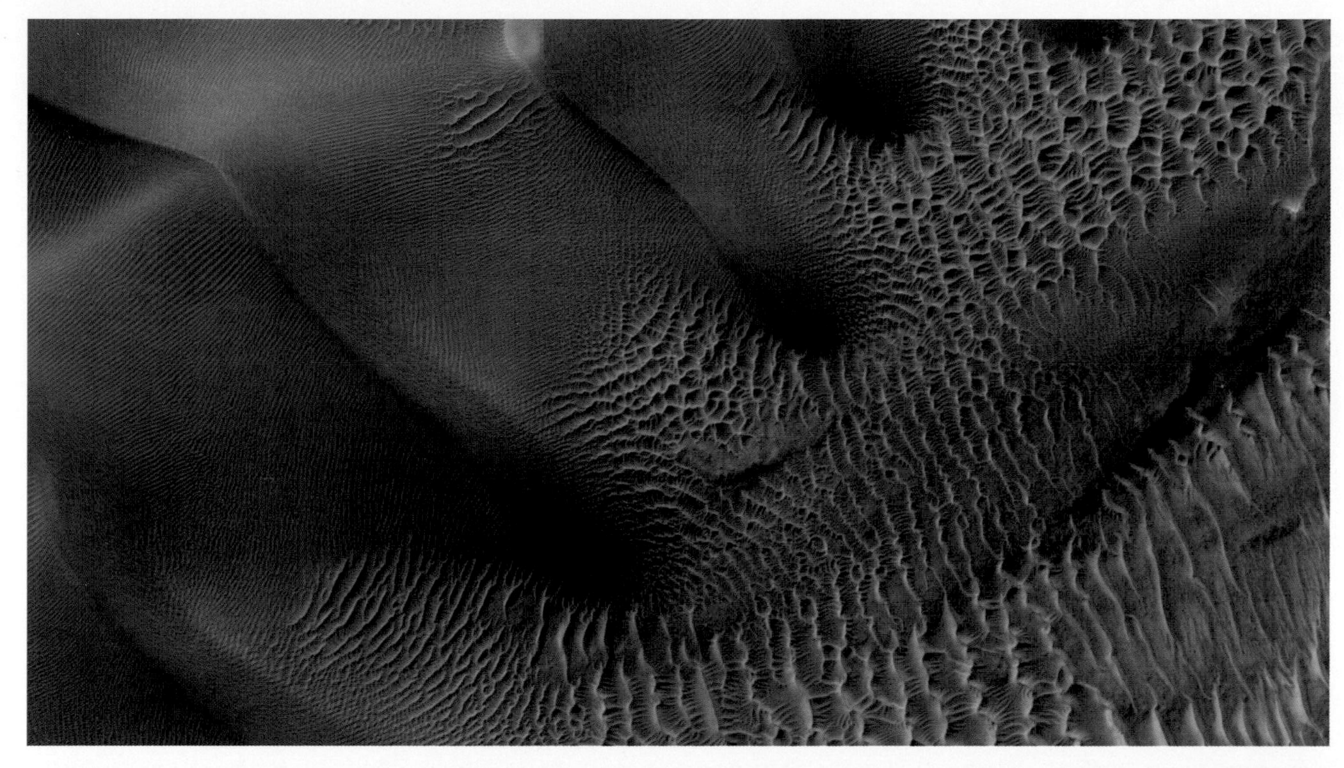

ALL:

**Sand dune formation on Mars**

The processes of sand dune formation are somewhat different to those on Earth and take much longer. Nonetheless the results can look similar to a terrestrial desert, with linear, barchan and star dunes. Stark transitions in topography and surface temperature seem to be key to creating the conditions for sand dunes on Mars. The image on the right from *Mars Reconnaissance Orbiter* shows sand dunes in the Noachis Terra region, whilst the Bagnold Dune Field (opposite) was photographed by the *Curiosity* rover.

ALL:

**Rivers**

No rivers flow over the surface of Mars today; the atmosphere is too thin and the temperature too cold for liquid water to be sustained on the surface. However, no fewer than 40,000 river valleys have been mapped, apparently carved by running water. Some are sinuous whilst others are dendritically branched with tributaries, suggesting they were fed by rainfall. The winding river channel on the left is on the floor of Lyot Crater, whilst the image on the right shows gullies apparently formed by water erosion in the Hellas Impact Basin.

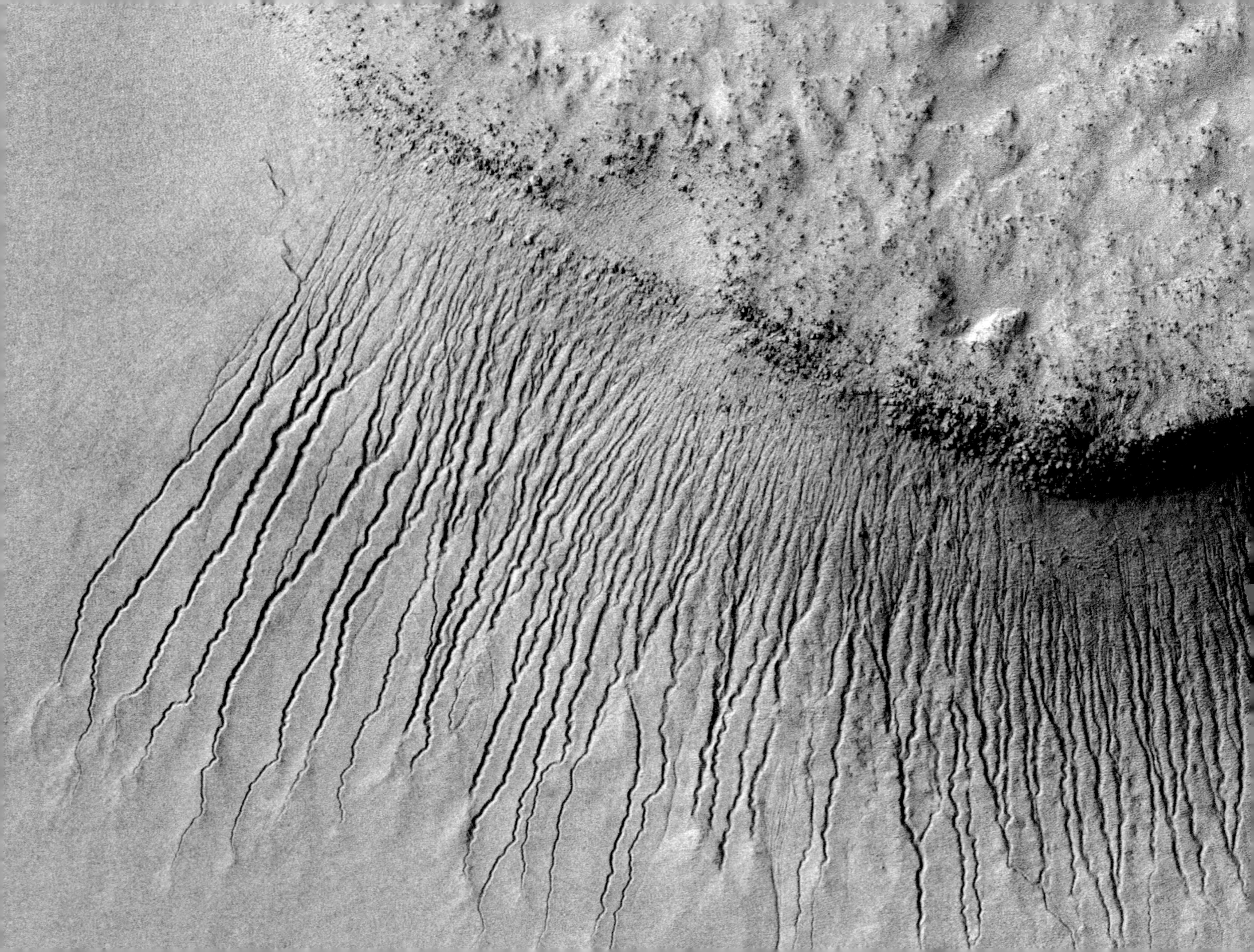

RIGHT:

**Deltas**

Complex deltas have been discovered where rivers flowed into former lakes, along with evidence of ancient shorelines. Some of these watercourses are clearly fossilized, the riverbeds now filled with later sediment which stands above the surrounding eroded topography. Many of these features are estimated to have been formed billions of years ago. This image shows a complex delta where an ancient river emptied into a former lake in Jezero Crater.

LEFT:

**Flood waters**

In the ancient past, flood waters scoured enormous outflow channels on the surface of Mars, such as here in Niger Valles.

OVERLEAF:

**Kasei Valles**

One of the largest outflow channels on Mars is Kasei Valles, a giant system of canyons up to 500km (300 miles) wide, shown in this perspective image by *Mars Express*.

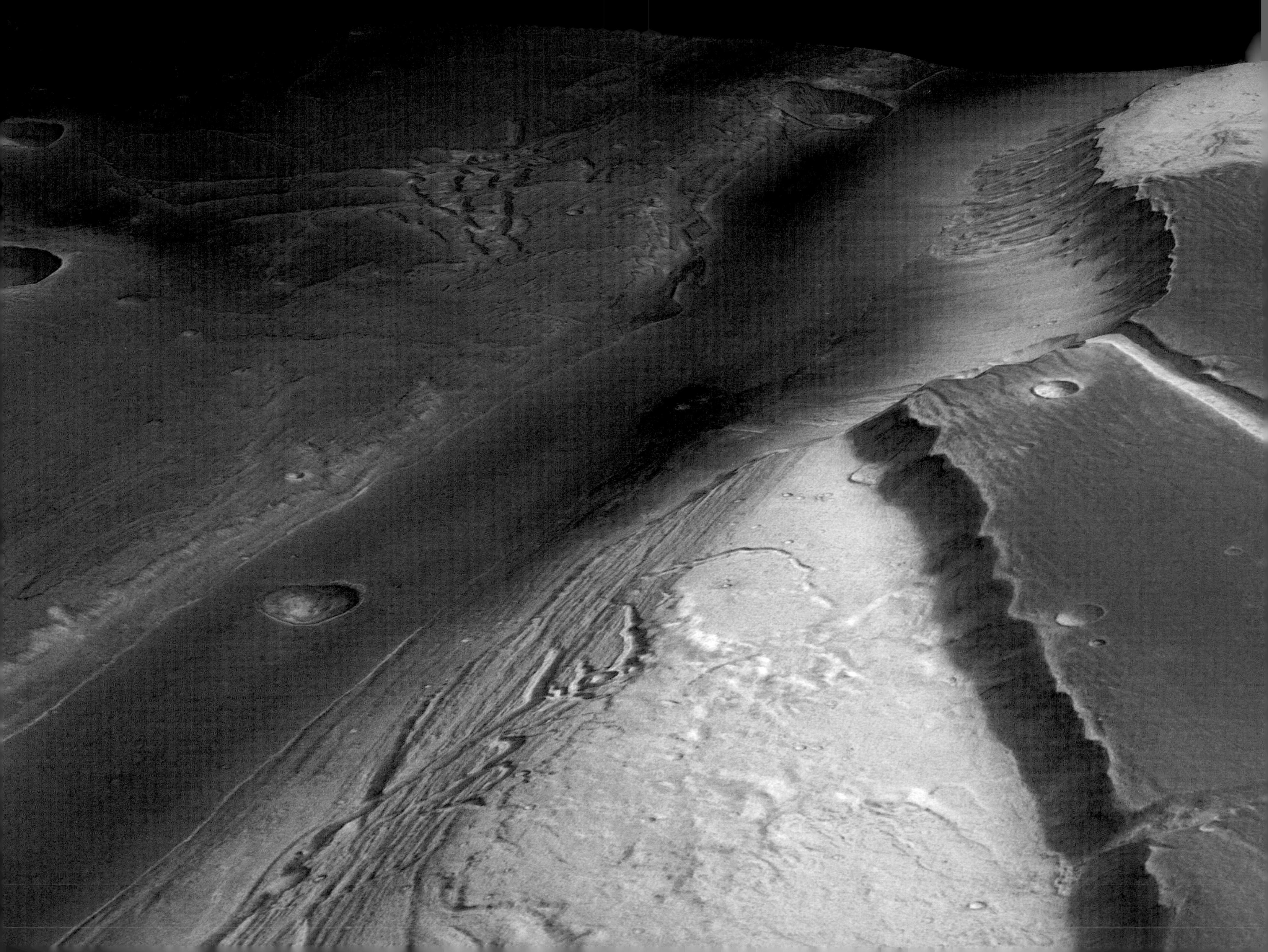

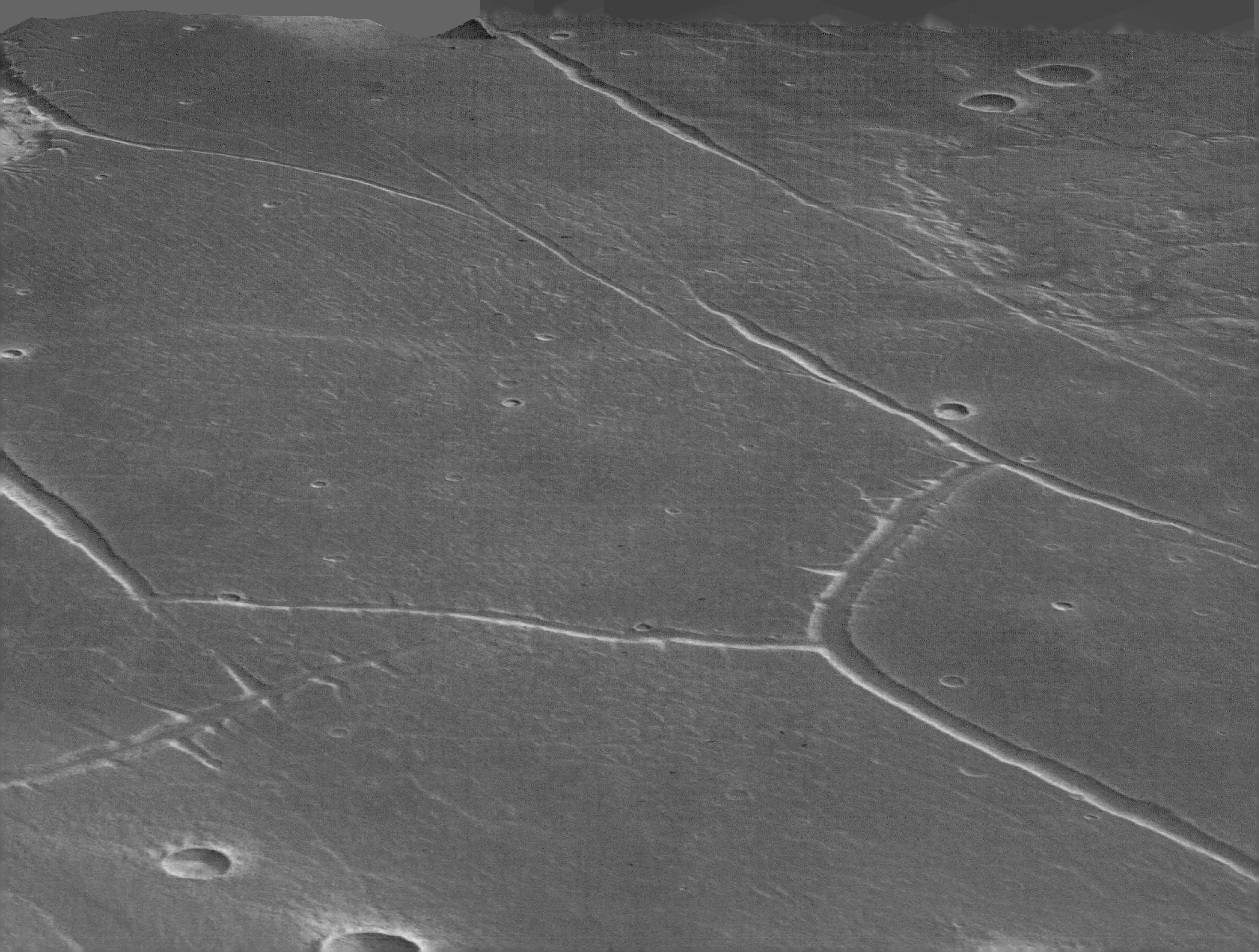

ALL:
**Valles Marineris**
One of the most striking features on the surface of Mars extends around 20 per cent of the planet's circumference, almost along the equator. Valles Marineris has the appearance of a huge canyon more than 4000km (2500 miles) long, 200km (120 miles) wide and up to 7km (4 miles) deep. The chasm is six times longer, six times deeper and 10 times wider than Earth's Grand Canyon, which would fit into it more than one hundred times. It was discovered by *Mariner 9*, after which it is named. Several theories have been proposed for the formation of Valles Marineris. The most widely accepted is that it originated by rifting of the Martian crust, making it more analogous to the Great Rift Valley in Africa rather than the Grand Canyon. As Mars lacks plate tectonics, the fracture is thought to have resulted from isostatic loading of the crust as volcanoes formed the massive Tharsis uplands. Flowing water and landslides subsequently widened the rift and created numerous tributaries. An area of chaotic terrain and outflow channels marks the eastern end of the Valle where it widens into the lowlands of Chryse Planitia.

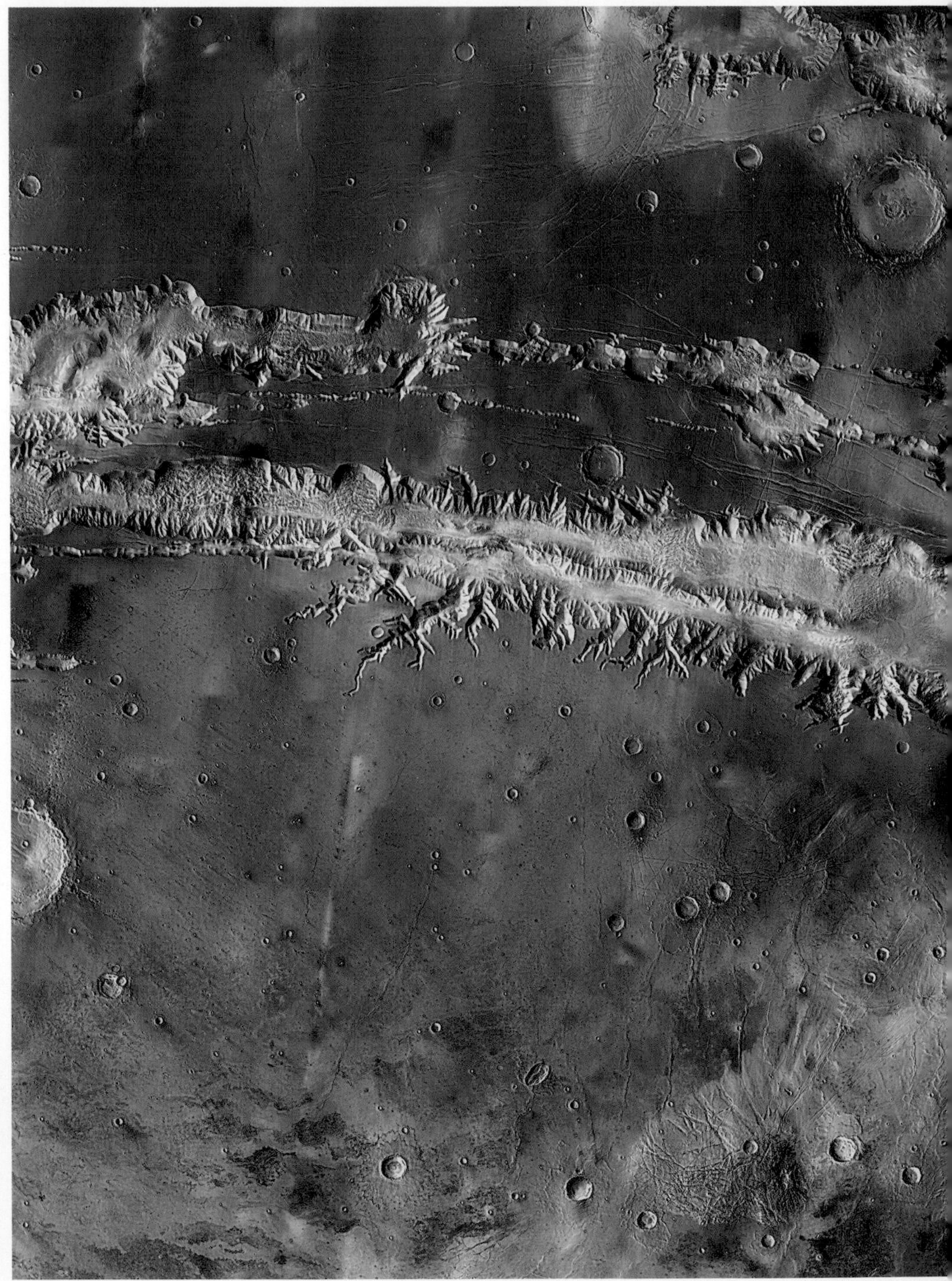

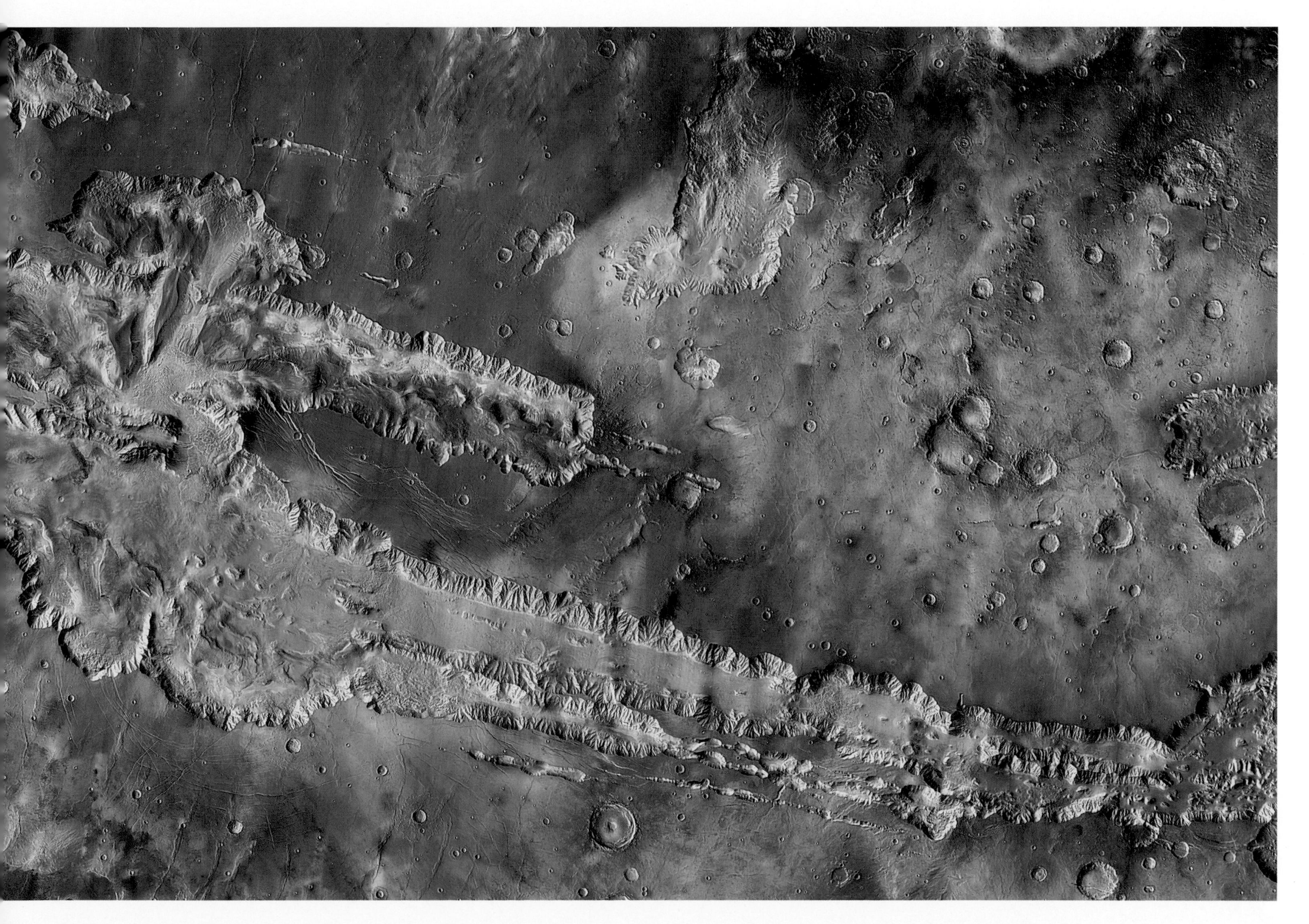

### Exploring the surface of Mars

Since 1997, six rovers have landed and driven on the surface of Mars. Their landing sites have been selected to investigate places where water once flowed. *Opportunity*, which landed in 2004, was operational for 5352 sols (Martian days), which is 15 Earth years, during which time it drove 46km (28 miles). Its sibling *Spirit* managed 2208 sols (six Earth years) and nearly 8km before becoming stuck in a sand trap. Real-time driving of rovers by humans on Earth is not possible as radio signals take several minutes to reach Mars. Instead, rovers are programmed in advance to operate autonomously.

*Curiosity* (right) and *Perseverance* are the most advanced rovers so far despatched to Mars. Each is about the size of a car and both are currently operational.

**Marathon Valley photographed by *Opportunity***
This panorama shows an area investigated by *Opportunity* for evidence of how ancient water flows altered rocks. It was selected for investigation following the discovery of water-related clay minerals by *Mars Reconnaissance Orbiter*.

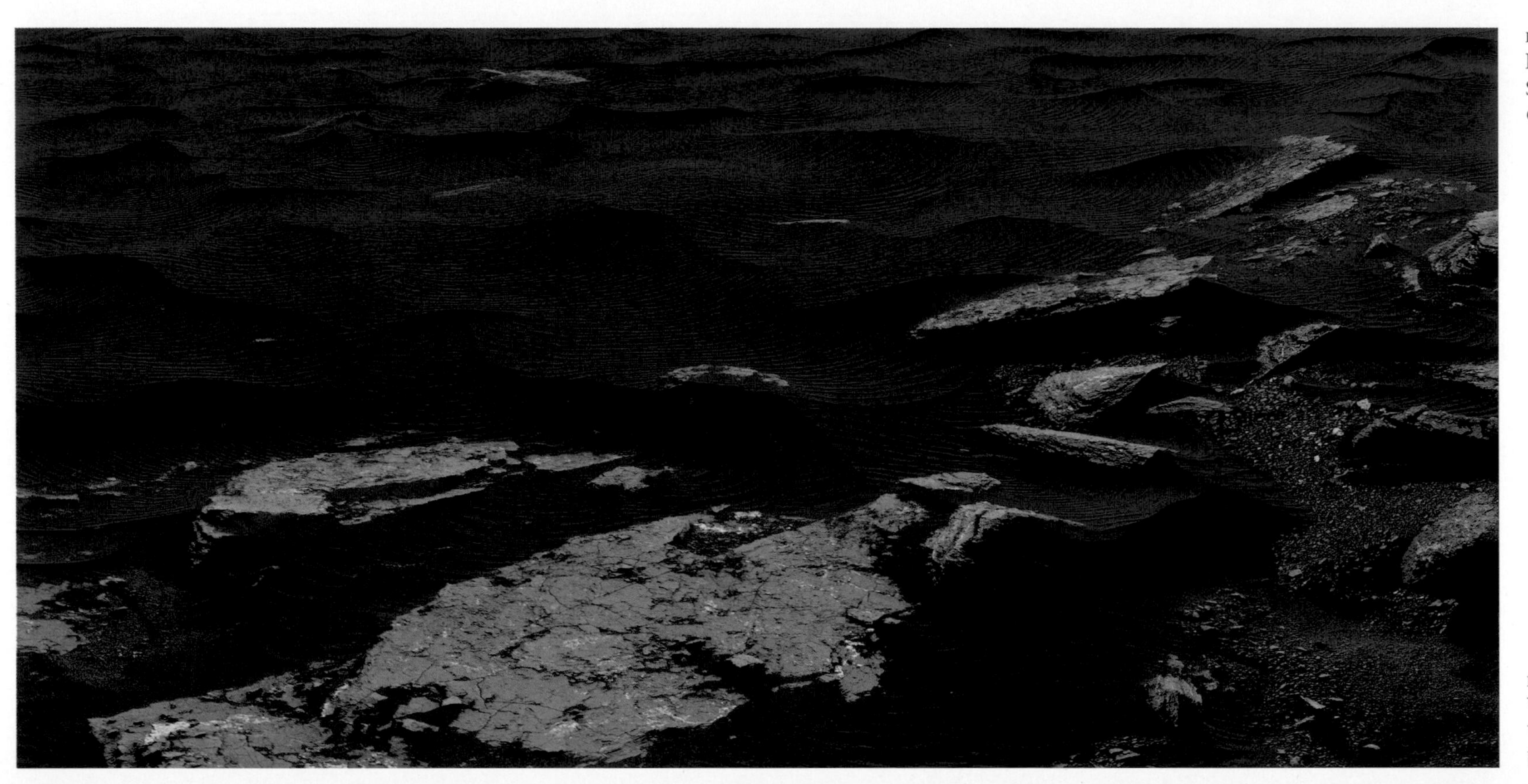

LEFT:
**Dunes on Mars**
Sand dunes photographed by *Curiosity*.

BELOW:
**Victoria Crater**
Victoria Crater photographed by *Opportunity*.

LEFT:

**Martian rocks in Gale Crater**

Curiosity landed in 2012 within Gale Crater, which is believed to have contained a lake three-and-a-half billion years ago. These rocks were investigated by *Curiosity* for possible fossil evidence of past life.

OPPOSITE:

**Landing *Perseverance***

*Perseverance* arrived on the surface of Mars on 18 February 2021. Like *Curiosity* in 2012, it was landed using extraordinary technology called a sky crane. Following entry into the atmosphere using a heat shield and parachutes, the rover was suspended beneath the spacecraft and lowered to the ground.

BELOW:

***Perseverance* and *Ingenuity***

The rover carried with it a small helicopter named *Ingenuity*, which on 19 April 2021 became the first powered aircraft to fly on another planet.

RIGHT:

***Perseverance* in Jezero Crater**

*Perseverance* is exploring Jezero Crater (pictured right), another likely ancient lake. Its landing site in 2021 was just a few kilometres from an ancient delta photographed by *Mars Reconnaissance Orbiter* (Page 126).

The surface of Mars has no liquid water, little atmosphere, no ozone layer to shield it from ultraviolet radiation and no magnetosphere to prevent bombardment by solar wind. It seems inhospitable for current life. A major objective of all six rovers has been to look for signs of past life, which may have existed billions of years ago when the planet was warmer and wetter. Each rover has been equipped to gather and analyse material from the Martian surface.

*Perseverance* is collecting rock, soil and gas samples into tubes which it is leaving on the surface of Jezero Crater for retrieval and return to Earth by a future mission.

RIGHT:

**Was Mars once a warm, wet planet?**

Given the abundant evidence of flowing water on its surface, many scientists believe that in the distant past Mars was very different to the desert world we see today. The spacecraft *Mars Odyssey* (in orbit since 2001) and *Phoenix* (which landed in 2008) detected the presence of large volumes of subsurface water ice. If all this were melted it would be sufficient to cover the whole planet to a depth of 35m (100ft). Radar imaging by *Mars Express* has also found evidence of a subglacial lake of liquid water beneath frozen deposits near the South Pole.

The Mars ocean hypothesis proposes that this water once formed an ocean in the northern hemisphere, covering about one-third of the planet to a depth of several kilometres. Evidence of ancient shorelines, 3.8 billion years old, has been traced around the perimeter of this vast basin, which is named Vastitas Borealis. There are even signs of boulders carried by tsunamis caused by asteroids striking the ocean.

If Mars did have such an ancient ocean, it must have had a thicker atmosphere at the time to create a greenhouse effect sufficient to raise temperatures above freezing. Most of this early atmosphere has subsequently disappeared. Because Mars is a small planet with lower gravity than Earth, gases such as hydrogen can readily escape to space. The lack of a planetary magnetic field exposes the Martian atmosphere to depletion by the solar wind, resulting in heavier gases such as carbon dioxide being lost. These are key reasons why today the Red Planet is cold and arid, whilst our own blue planet is warm and wet.

Pictured right is an artist's impression of how Mars may have looked 3.8 billion years ago.

RIGHT:
**Phobos**
Mars has two moons, Phobos and Deimos, both of which are tiny compared to Earth's Moon. As they have little gravity of their own they are irregular in shape. Their density is too low to be solid rock. More likely they are piles of rubble loosely held together by weak gravity.

Phobos is 27km (16.7 miles) across its widest axis and orbits just 6000km (3730 miles) above the Martian surface. Since it takes only 7 hours 39 minutes to complete an orbit, Phobos appears to rise in the west, cross the Martian sky in a few hours and set in the east. Annular solar eclipses can be seen from the surface of Mars when Phobos transits the face of the Sun and have even been photographed by rovers. Because it orbits in less than one Martian day, tidal interactions are dragging Phobos closer to Mars, so that within 50 million years it will break up to form a planetary ring.

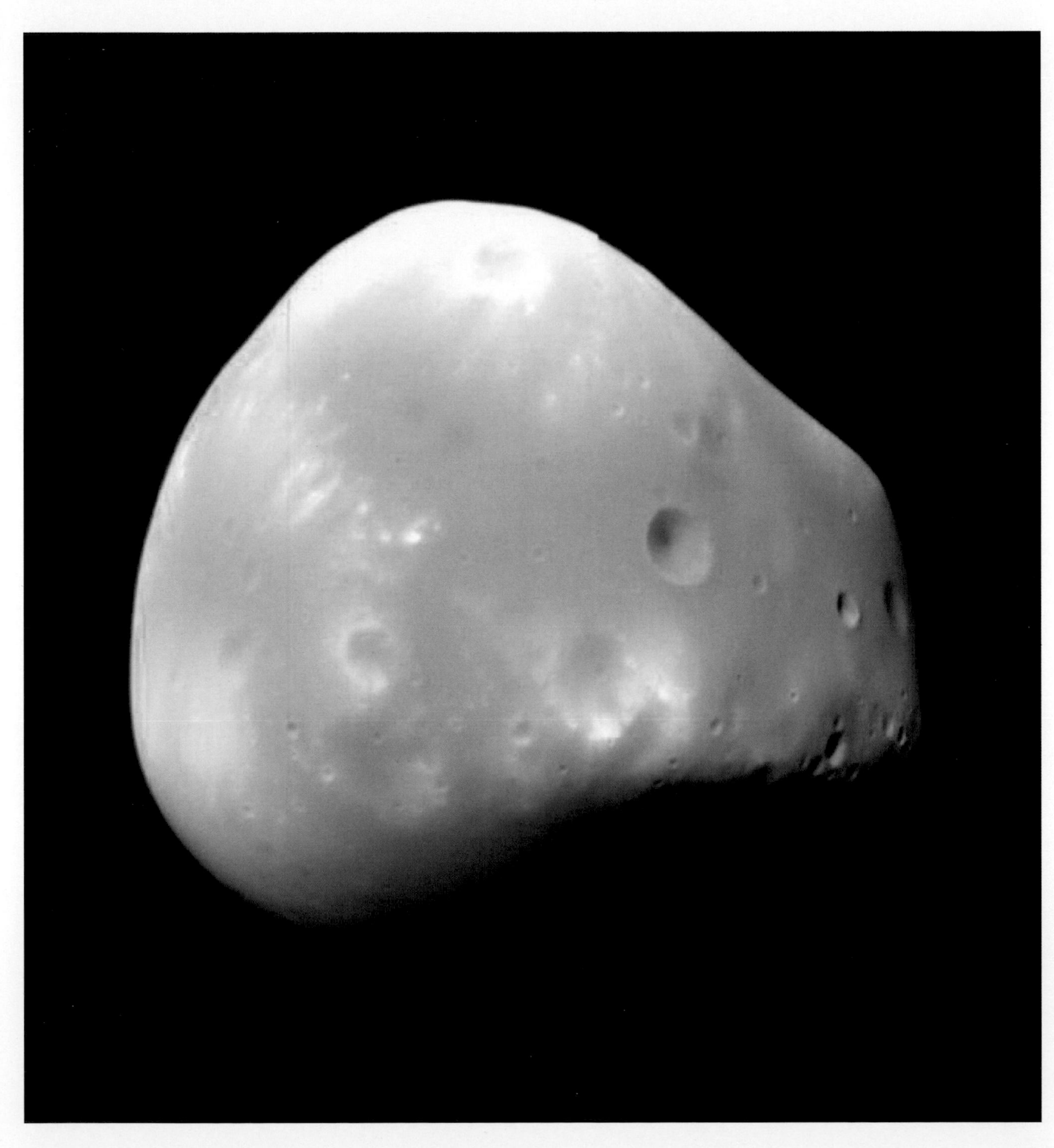

LEFT:
**Deimos**
With a maximum width of just 15km (10 miles), Deimos is half the size of Phobos and 23,000km (14,000 miles) from Mars. Its surface is smoother and less cratered than that of its sibling. Deimos completes each orbit in 30 hours, a little longer than Mars takes to rotate so it appears to rise in the east and set in the west. When fully illuminated it appears in the Martian sky about as bright as Venus does from Earth.

The origin of Phobos and Deimos is uncertain. They could be captured asteroids or debris from a collision of a large object with Mars early in its history. Another theory is that they could be fragments of a single larger moon that broke up as a result of an asteroid impact.

# Jupiter and its Moons

Named by the Romans after the king of their gods, Jupiter is undoubtedly king of the planets. This gas giant has two and a half times the mass of all the other planets combined.

With a diameter of 143,000km (89,000 miles), Jupiter is 318 times heavier than our own planet and could fit over 1000 Earths inside its volume. Like the Sun, it is mainly composed of hydrogen and helium. These gases become increasingly dense within the planet until hydrogen forms what is effectively a liquid. The mantle comprises metallic hydrogen, which gradually defuses into what is thought to be a rocky core of heavier elements.

Giant planets have a thick atmosphere but no solid surface beneath. They formed further from the Sun than terrestrial planets where temperatures are lower, enabling volatile components to condense into liquids and solids. Jupiter is a turbulent planet with strong winds, giant storms and an intense magnetic field. Although too small to be a star, it generates more heat internally than it receives from the Sun.

This huge planet lies at the centre of a Jovian System of smaller bodies in orbit around it. To date, 80 natural satellites have been discovered, most of which are less than 10km (6 miles) in diameter. The four largest, however, are amongst the six biggest moons in the Solar System. Io, Europa, Ganymede and Callisto were discovered by the astronomer Galileo when he pointed his telescope to the heavens in 1609 and 1610. At the time, this provided irrefutable proof to counter the orthodox idea that everything in the Universe revolves around Earth.

Since 1973, nine spacecraft have visited Jupiter, of which two have orbited around it for several years. They have sent us stunning pictures of the planet itself and made fascinating discoveries about the enigmatic worlds in orbit around it.

OPPOSITE:

**Jupiter**

Jupiter's stripes and swirls are clouds of ammonia crystals at the top of its atmosphere, which is predominantly hydrogen and helium. Seen here in images captured by the *Juno* spacecraft, the clouds occur as bands at different latitudes, referred to as zones (light-coloured) and belts (dark hues). The clouds are driven by winds of up to 600km/h (370mph), creating turbulent circulation patterns and storms.

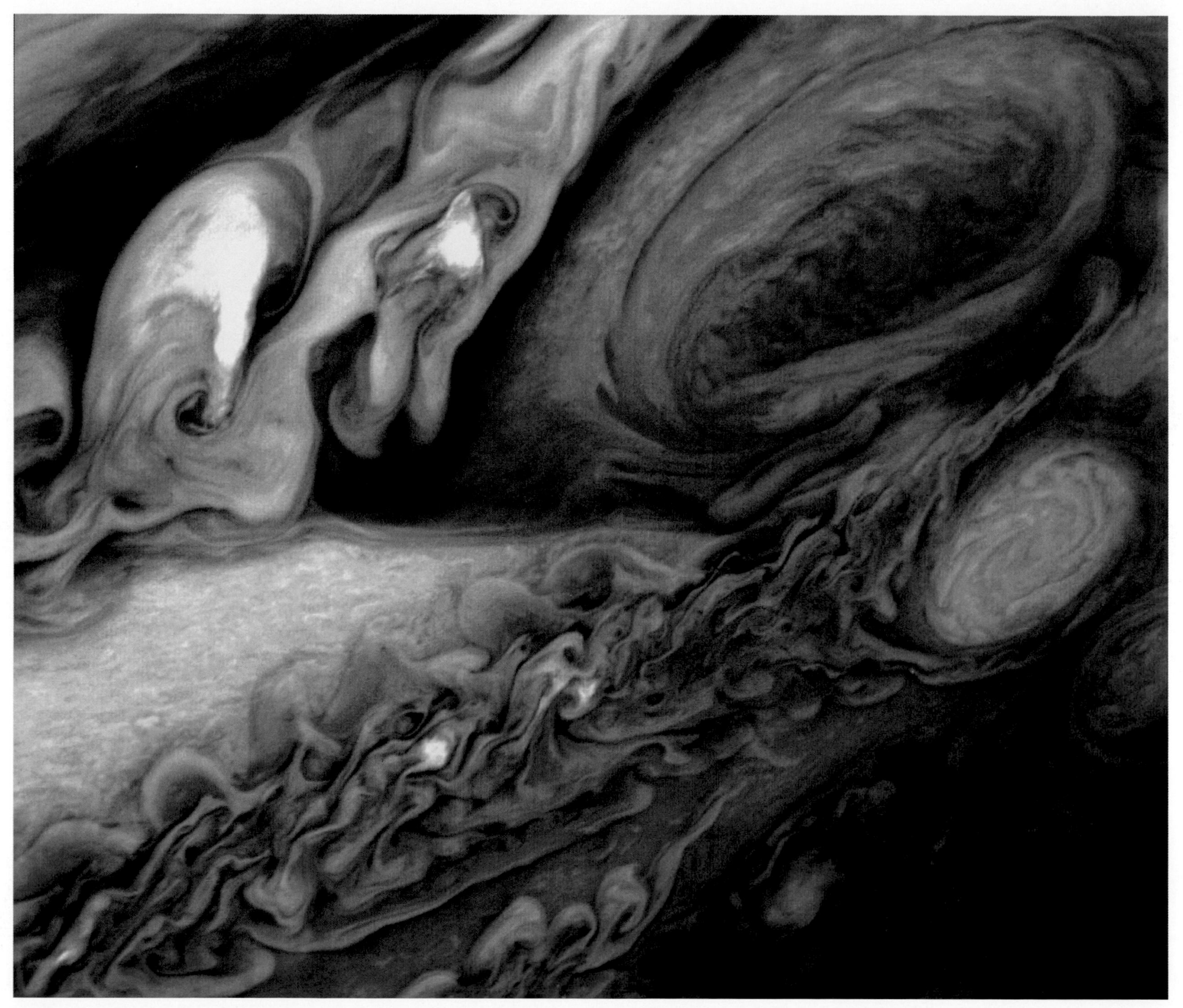

LEFT:
**Great Red Spot**
The best-known feature on Jupiter is its iconic Great Red Spot, a giant storm whose diameter is bigger than Earth. It was observed through telescopes in the nineteenth century and possibly as long ago as 1665, so it is known to have raged for centuries. In contrast to major storms on Earth which are cyclones (low pressure), the Great Red Spot is an anticyclone (high pressure). As it is located in Jupiter's southern hemisphere its rotation is anticlockwise, taking six days to complete. The distinctive red colour may be the result of ammonia reacting with acetylene. This image was captured by *Voyager 1* as it passed the planet in 1979.

OPPOSITE:
**A Jovian day**
Jupiter rotates on its axis every 9 hours 56 minutes, the fastest of any planet. The resulting centrifugal force creates an equatorial bulge so Jupiter's diameter is noticeably wider across the equator (143,000km/89,000 miles) than the poles (134,000km/83,000 miles), as shown in this view by the *Hubble Space Telescope*. A Jovian year, on the other hand, is much statelier, taking 11.9 Earth years to complete.

LEFT:

**Jupiter's south pole**

This view of the planet's southern hemisphere was captured by *Juno* as it passed over Jupiter's south pole in December 2017. Colourful cloud features include parallel reddish-brown and white bands, multi-lobed chaotic regions, white ovals and many small vortexes. Many clouds appear in streaks and waves due to continual stretching and folding by Jupiter's winds and turbulence. *Juno* discovered a cyclone at the south pole with five more cyclones around it, each around 4500km (2800 miles) in diameter with wind speeds of 360km/h (220mph).

OPPOSITE:

**Jupiter's magnetic field**

Jupiter has the most intense magnetic field of any planet, 14 times stronger than Earth's. The resulting magnetosphere extends hundreds of millions of kilometres into space, almost as far as Saturn. In much the same way as on Earth, the incoming solar wind is intercepted by the planet's magnetic field and charged particles are funnelled towards the poles. This results in spectacular aurorae at high latitudes, photographed here by *Hubble Space Telescope*.

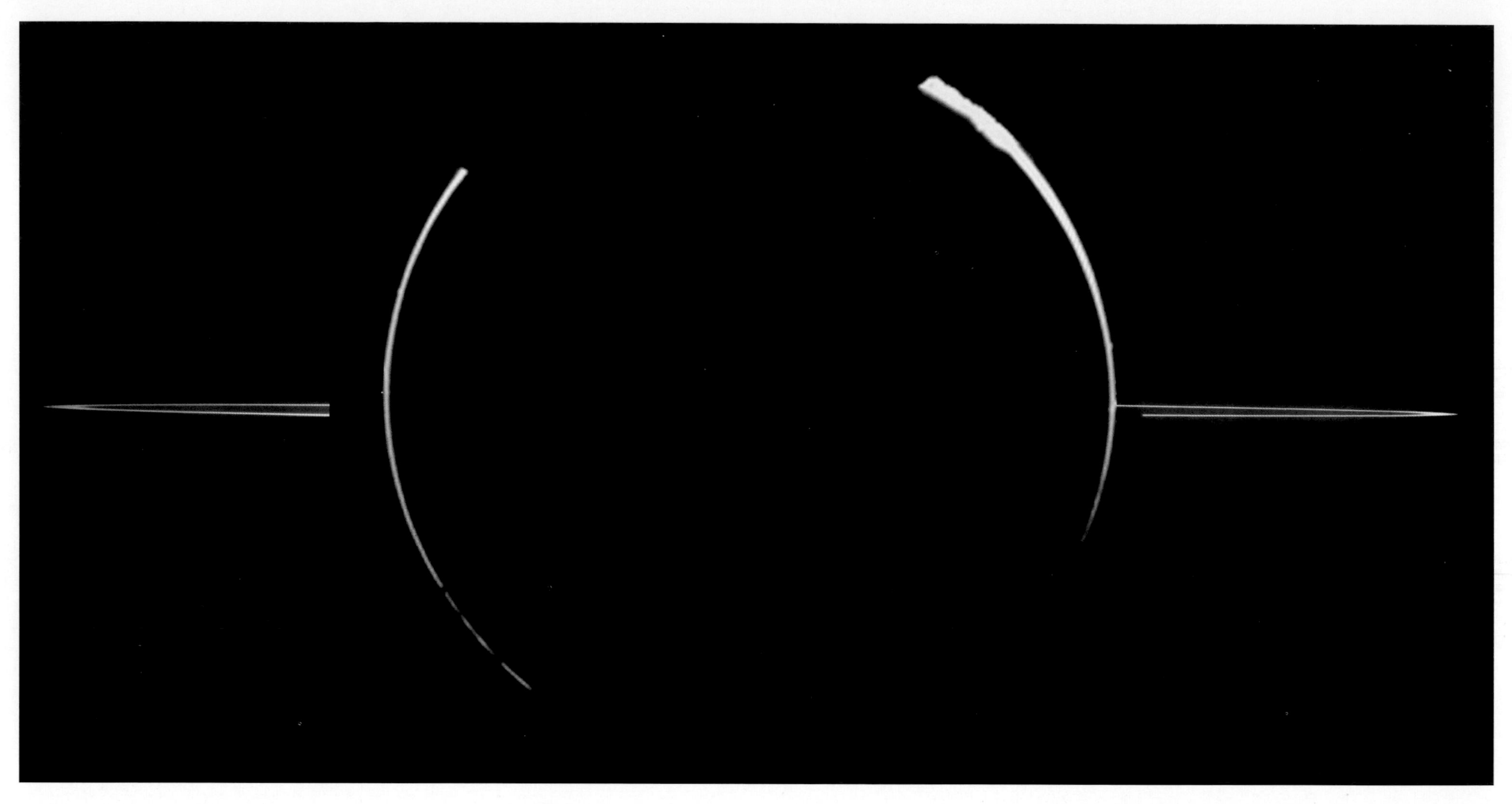

ABOVE:
**Jupiter's rings**
Although Saturn is best known for its rings, in fact all four giant planets have ring systems. Those of Jupiter are faint and composed mostly of dust originating from the tiny moons Metis and Adrastea. There are four of them: halo ring, main ring and two gossamer rings associated with the moons Amalthea and Thebe. The rings were discovered by *Voyager 1* in 1979. This photograph by *Galileo* shows Jupiter and its rings backlit as the probe passed behind the night side of the planet.

OPPOSITE:
**Visiting Jupiter**
The first spacecraft to fly-by Jupiter was *Pioneer 10* in December 1973. It was followed by *Pioneer 11* in December 1974, *Voyager 1* in March 1979 and *Voyager 2* in July 1979, each of which used Jupiter's huge gravity to boost their speed and continue to Saturn. The next stage in studying Jupiter was to send a spacecraft to orbit the planet. This was achieved by *Galileo* (shown in the artist's impression opposite) between December 1995 and September 2003. During more than seven years, *Galileo* made 35 orbits of the planet, enabling multiple fly-bys of all four Galilean moons. It also released a probe into the Jovian atmosphere which parachuted to a depth of 150km (90 miles). The probe returned data for almost an hour before it was crushed by high pressure and temperature.

*Galileo* was succeeded by *Juno*, which entered orbit in July 2016 to study Jupiter's composition, atmosphere, gravitational and magnetic fields. Research objectives include seeking evidence of the nature of Jupiter's core and how the planet formed. *Juno* has a polar orbit, enabling views of high latitudes in much more detail than previous missions but not optimised for close observation of the moons. Jupiter was also visited and photographed by *Cassini* in 2000 on its way to Saturn and *New Horizons* in 2007 en route to Pluto.

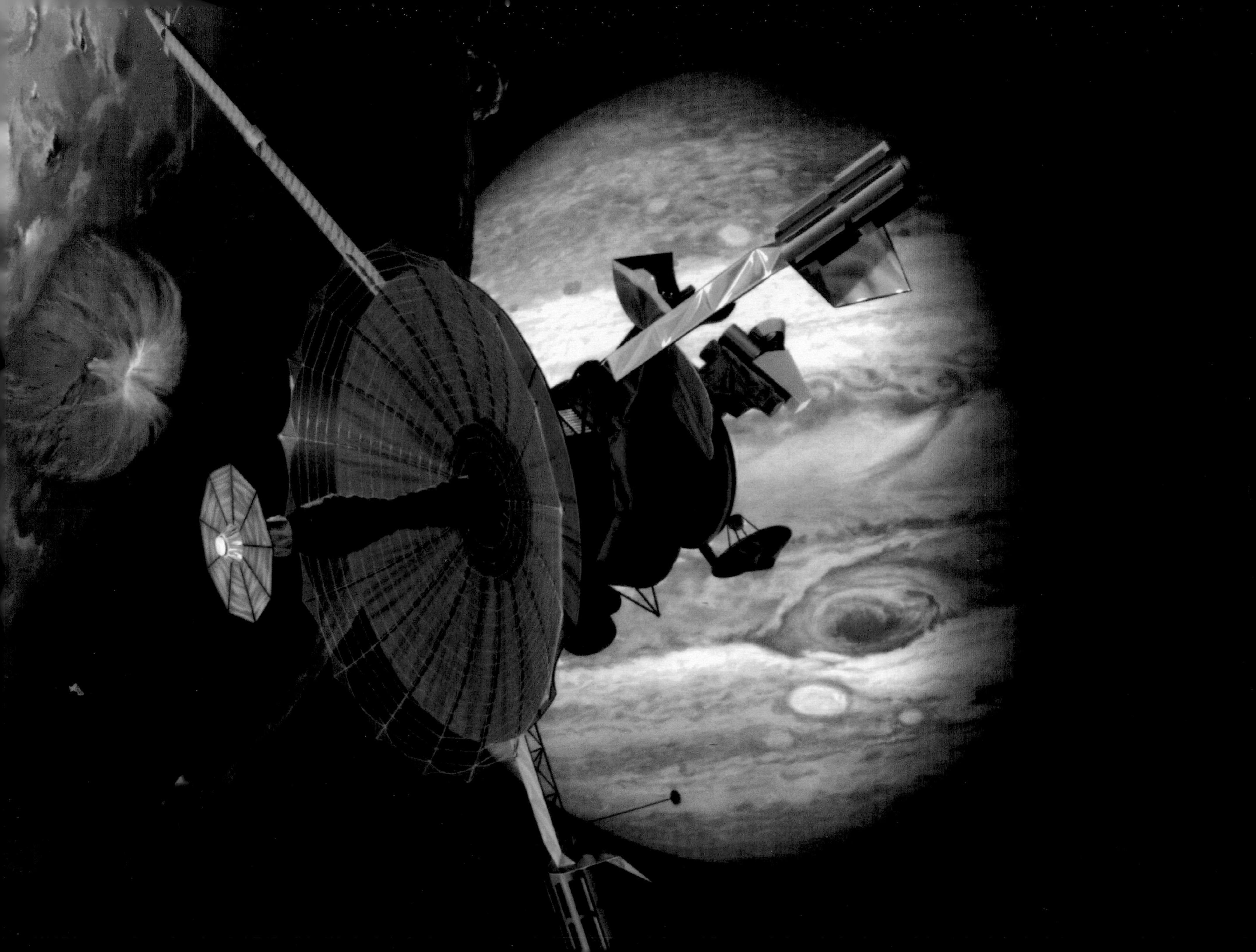

LEFT:

**Galilean moons**

The four largest moons of Jupiter discovered by Galileo in 1610 are fascinating worlds. The innermost, Io and Europa, are similar in size to Earth's Moon. Ganymede and Callisto, which orbit further out, are significantly larger. This composite shows the relative sizes (but not the relative locations) of Io, Europa, Ganymede and Callisto in comparison to Jupiter's Great Red Spot. The four Galilean moons were named for the mythological lovers of Zeus, the ancient Greek equivalent of Jupiter.

The gravitational tugs of the moons on each other have resulted in orbital resonance between three of them. Io orbits in 1.8 days, Europa in 3.6 days and Ganymede in 7.2 days. So for every orbit completed by Ganymede, Europa makes exactly two and Io makes exactly four. Callisto does not partake in this cosmic formation dance, taking 16.7 days to circuit Jupiter.

All four Galilean moons are tidally locked to Jupiter so they always present the same face to the planet (as our Moon does to Earth). Being different distances from Jupiter, each moon experiences a different degree of tidal heating, as a result of which they are radically different from each other.

OPPOSITE & RIGHT:

**Io**

One of the most remarkable discoveries of the *Voyager* spacecraft is that Io has active volcanoes. This was first realized when a mission researcher noticed a plume emanating from the surface. *Voyager 1* revealed a strange, multi-coloured landscape that lacks impact craters and instead is pock-marked with volcanic eruptions. Io is the most geologically active world in the Solar System, with a surface constantly renewed by lava flows. Its bizarre colour results from a frosty coating of sulphur belched from 400 active volcanoes. There are over 100 mountains on Io, some of them taller than Mount Everest.

Io's interior remains molten as a result of tidal heating. In essence, Io is squeezed by Jupiter's intense gravity, giving rise to internal forces that are 20,000 times stronger than Earth experiences from tidal interactions with our Moon. These forces would dissipate over geological time (as have those between Earth and Moon), were it not for Io's orbital resonances with Europa and Ganymede. These resonances keep Io confined in an eccentric orbit just 350,000km (220,000 miles) above the Jovian cloud tops.

Not only is Io pummelled by Jupiter's gravity, it is also bombarded with intense radiation and magnetism from the giant planet. In fact, the moon plays a significant role in shaping Jupiter's whole magnetosphere. Dust and gas from Io's volcanoes are swept into the Jovian magnetic field, whilst the whole moon acts as an electric generator with potential energy of up to 400,000 volts. These extreme conditions make Io a challenging and hazardous environment for visiting spacecraft, let alone any kind of life.

The photograph to the right shows the extraordinary world of Io, coloured yellow, orange, red, white and black by sulphurous compounds emitted by 400 active volcanoes.

Volcanic plumes on Io are pictured opposite. The caldera Pillan Patera is erupting on the edge of the moon, producing a plume 140km (87 miles) high. A plume from Prometheus, near the centre of the disc, is producing a reddish shadow. Prometheus was observed erupting by both *Voyager* in 1979 and *Galileo* in 1997.

RIGHT:
**Europa**
In contrast to the rugged mountains and volcanic calderas of Io, the smallest of the Galilean moons looks like a billiard ball. Europa is a smooth sphere of ice.

With a temperature of -160°C (-256°F) at its equator, Europa's surface ice is frozen as hard as rock. However, several lines of evidence indicate that there is an ocean of liquid water beneath the ice. The ocean is kept from freezing through heat generated by tidal forces as Europa orbits Jupiter. This is the same mechanism that drives Io's volcanoes but to a smaller degree, as Europa orbits further from the giant planet. The moon is 600,000km (370,000 miles) from the Jovian cloud tops and has an almost circular orbit, so the stretching and squeezing by Jupiter's gravity is less intense than for Io.

The thickness of Europa's ice is not known. Scientists' best estimates are that the ice layer extends between 10km and 30km (6–20 miles), below which lies an ocean of salty water 100km (60 miles) deep, overlying a rocky mantle. If that is correct, Europa has two to three times more liquid water than Earth. The likely existence of liquid water on Europa makes it a prime candidate in the search for extraterrestrial life. Beneath kilometres of ice, there is no sunlight to drive food chains. However, geothermal activity may occur on the ocean floor and could provide the necessary energy. Deep in Earth's ocean where sunlight cannot penetrate, hydrothermal vents create such conditions and sustain life independently of photosynthesis.

Several proposals have been put forward for spacecraft that could visit Europa and search for signs of life using ice-penetrating radar, direct impact onto its surface or even drilling through ice. *Europa Clipper* is scheduled for launch in 2024, to enter orbit around Jupiter in 2030 and make repeated fly-bys of the icy moon to assess its habitability and the prospects for landing a future probe. It could fly through plumes of water vapour, enabling the subsurface ocean to be sampled without the need to land and drill through the icy crust.

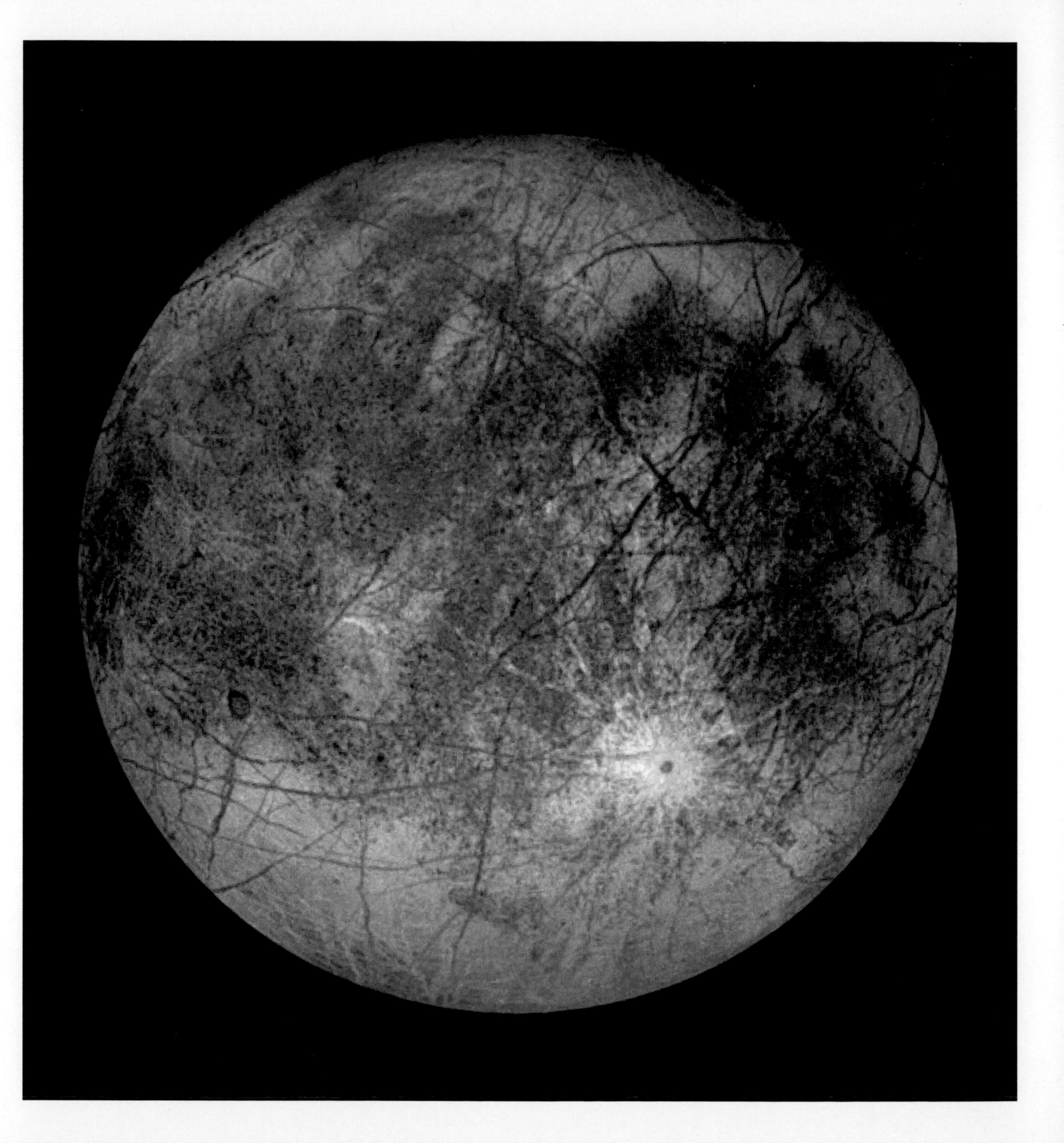

LEFT:
**Europa's lineae**
Reddish-brown lineae are the result of warmer ice rich in minerals (possibly sulphate) extruding from below.

Higher resolution images show cracks and ridges in the ice. The surface has few craters, indicating that it is relatively young. Tidal flexing by Jupiter appears to cause plates of ice to fracture and move relative to each other in a process similar to terrestrial plate tectonics. Cracks in the ice are analogous to Earth's mid-ocean ridges; where plates of ice are moving apart, to be recycled into the interior at subduction zones. Data from both *Galileo* and *Hubble Space Telescope* suggest that plumes of liquid water may erupt periodically through the ice into space.

ALL:

**Ganymede**

With a diameter of 5269km (3300 miles), Ganymede is the largest moon in the Solar System. It is bigger than the planet Mercury, though less than half as massive because it mainly comprises silicate rock and water ice, with only a small iron core. In common with Mercury, Ganymede generates its own magnetic field, the only moon known to do so. A saltwater ocean is believed to exist beneath Ganymede's icy surface. Orbiting Jupiter at a distance of one million km (620,000 miles), tidal heating is weaker than on Europa so the frozen crust is thicker, estimated at around 200km (120 miles). Below this depth, there may be several layers of ice and liquid water above a solid rocky mantle surrounding the core. As with Europa, there is a theoretical possibility of life in these subsurface oceans, though being much deeper under ice the oceans of Ganymede would be a great deal harder to investigate.

About two-thirds of Ganymede's surface comprises light-coloured icy regions, whilst the remainder is dark cratered terrain. The darker regions contain clays and organic materials and are much older, probably formed around four billion years ago. The lighter surface exhibits grooves and ridges that may result from tectonic activity, indicating a degree of resurfacing that has erased ancient craters. This giant moon has polar caps of water frost, thought to be formed by the bombardment of ice by charged particles travelling along its magnetic field.

Ganymede will be further investigated by *Jupiter Icy Moons Explorer* (*Juice*), due to launch in 2023 and enter orbit in 2032.

Ganymede's icy terrain can be seen in the image below, covering a width of 162km (100 miles). A caldera-like feature with a boundary escarpment in the centre was possibly formed by volcanic flows of liquid water and slush leading to the collapse of surface material.

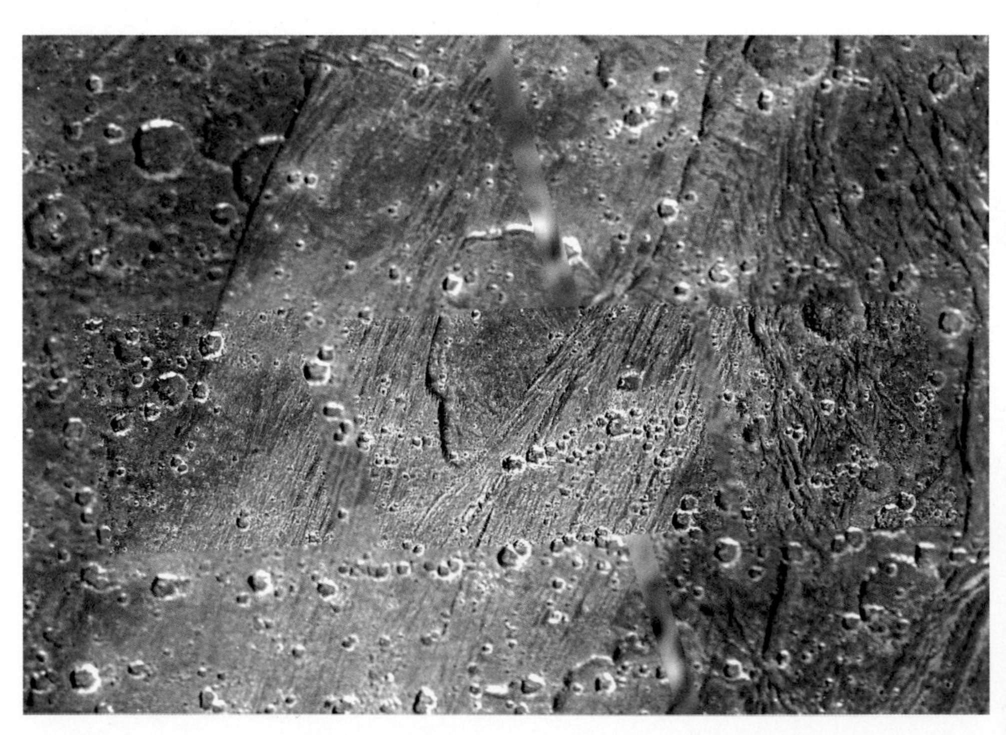

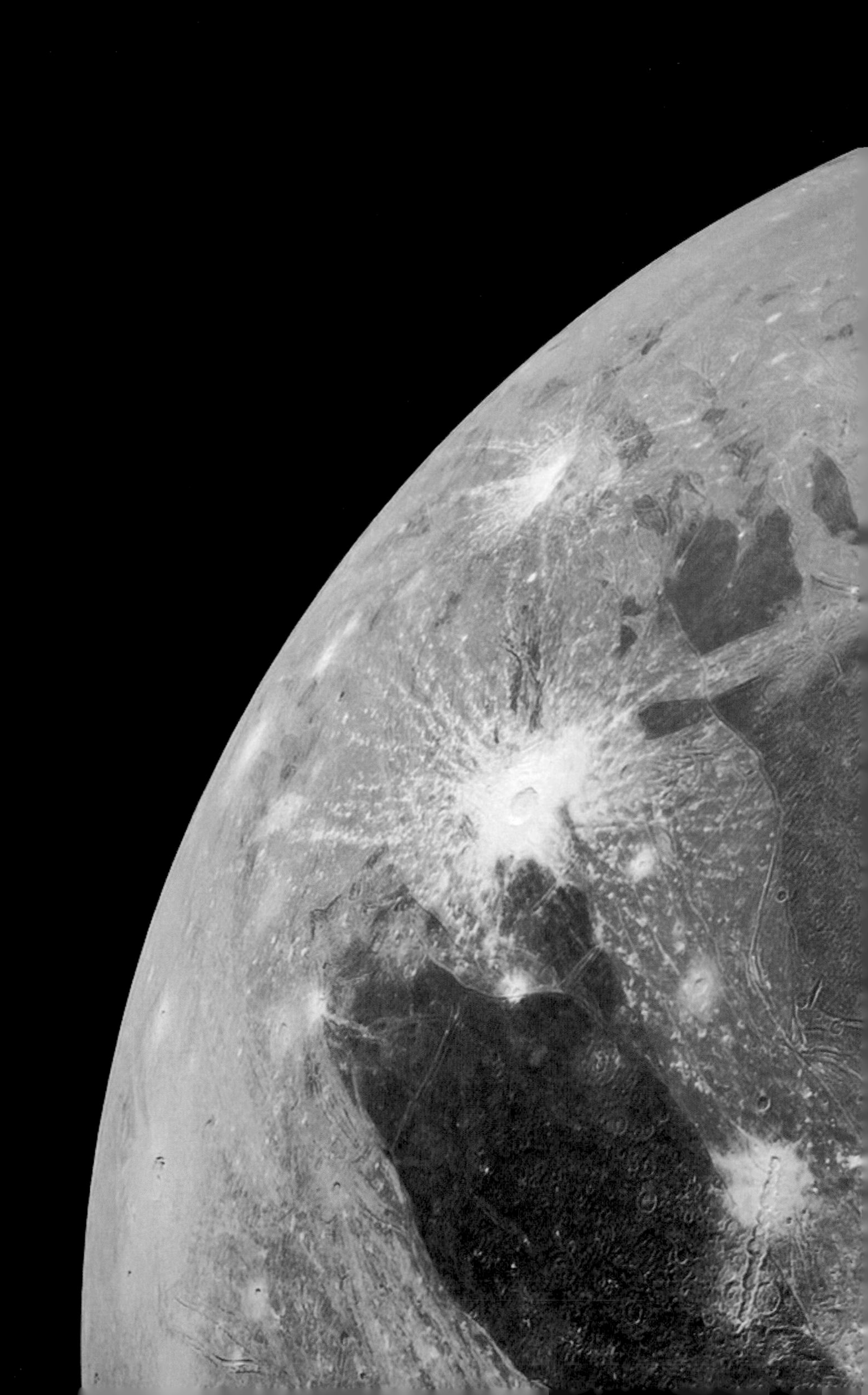

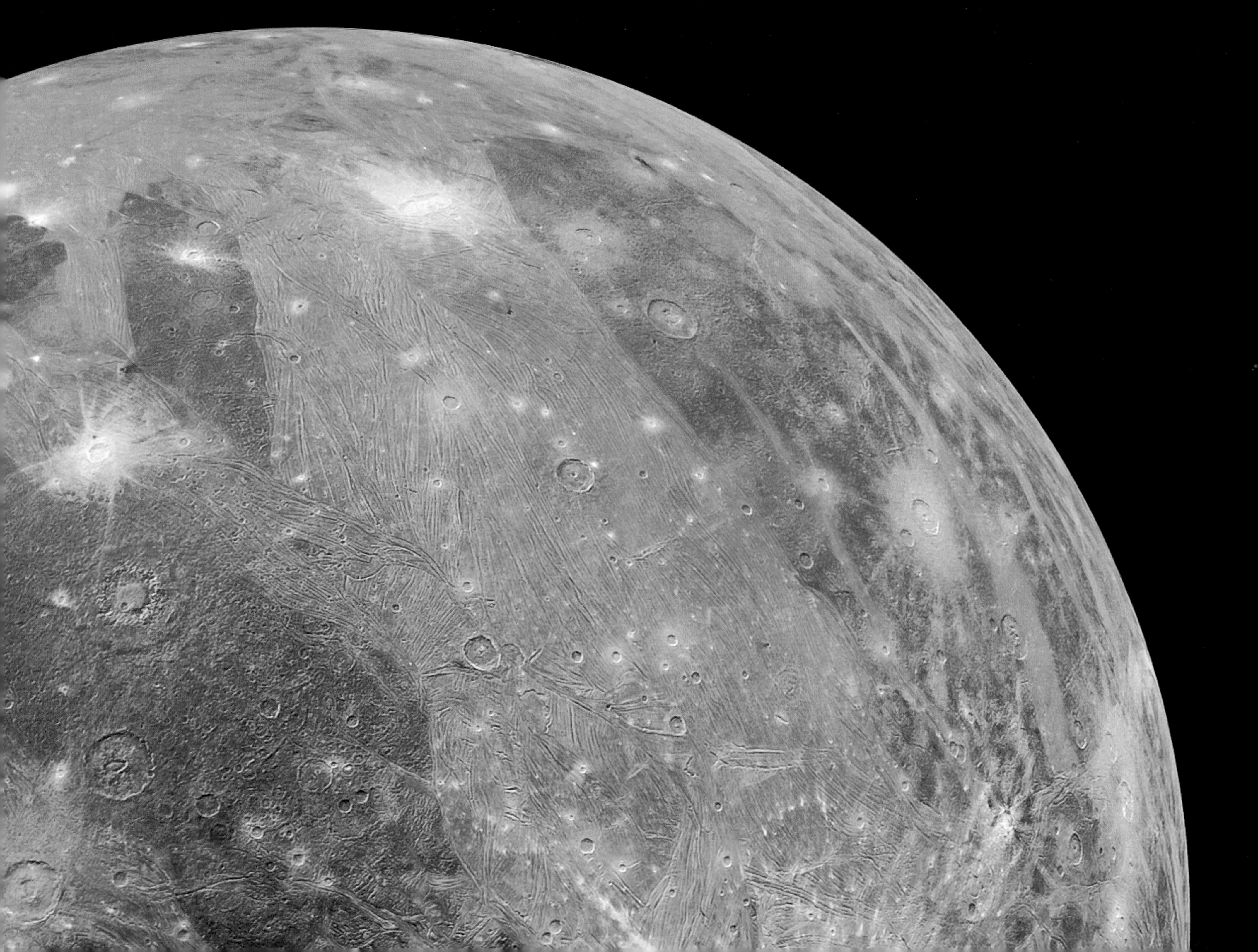

# Saturn and its Moons

Best known for its dazzling rings, Saturn is perhaps the most beautiful planet in the Solar System. Thanks to those rings and an incredible diversity of moons, it also has a good claim to be the most complex. In the last two decades, our knowledge and understanding of the Ringed Planet have been revolutionized thanks to an extraordinary 13-year mission of exploration by the *Cassini* space probe.

Like Jupiter, Saturn is a gas giant. It has no definite surface, though there may be a core of rock, iron and nickel. The atmosphere gets increasingly dense within the planet to form a deep layer of metallic hydrogen surrounding the core. However, most of the planet is gaseous, and Saturn is the only planet that is less dense than water. It is 95 times more massive than Earth, and its volume could contain 700 Earths.

Saturn's rings were first seen in 1610 by the astronomer Galileo, who thought he was looking at a triple planet. Their true nature was identified in 1658 by Christiaan Huygens, who also discovered Saturn's largest moon. It was later realized that the rings could not be a solid disc but must comprise numerous small particles, each orbiting the planet independently. We now know the rings comprise an estimated 30 million billion ($3 \times 10^{16}$) particles of water ice, making them extremely reflective.

Amongst Saturn's numerous moons, the two most fascinating are Titan, which has seas of liquid methane, and Enceladus, which is thought to have a subsurface ocean of liquid water. Titan is 200°C (392°F) colder than Earth but its thick nitrogen atmosphere, seas, rivers and rain make it in several ways the most Earth-like world in the Solar System.

OPPOSITE:

**Saturn and its rings**

Saturn rotates on its axis in around ten-and-a-half hours. This fast spin creates significant centrifugal force at the equator, which bulges outwards. This gives Saturn a noticeably flattened appearance, described as an oblate spheroid. The planet's equatorial diameter is 120,500km (74,800 miles) whilst from pole to pole measures only 108,700km (67,500 miles), which is 10 per cent less.

The first spacecraft to visit Saturn were *Pioneer 11* in September 1979, *Voyager 1* in November 1980 and *Voyager 2* in August 1981. Each zoomed past the planet at high speed, with observations and photographs acquired in just a few days.

LEFT:

**The composition of Saturn**

Like Jupiter, Saturn is made up mostly of hydrogen (96 per cent), with some helium (3 per cent). However, Saturn's atmosphere is relatively bland. Winds in the upper atmosphere reach speeds of 1800km/h (1100mph). These combined with heat rising from deep in the interior cause yellow and gold coloured bands at different latitudes. Metallic hydrogen deep within the planet generates an internal magnetic field. Although only 1/20$^{th}$ as strong as Jupiter's and weaker than Earth's, it is sufficient to generate spectacular aurorae around the magnetic poles.

OPPOSITE:

**Orbiting Saturn**

After a seven-year journey from launch, via fly-bys of Venus, Earth and Jupiter, on 1 July 2004 *Cassini* entered orbit around Saturn (artist's impression). Over the following 13 years, it became one of the most spectacularly successful missions of interplanetary exploration ever accomplished. Operating 1.4 billion kilometres from the Sun, *Cassini* could not use solar panels as its energy source. Instead, the probe was powered by the radioactive decay of 33kg (73lbs) of plutonium-238 carried on board, which generated sufficient electricity for 20 years' operation.

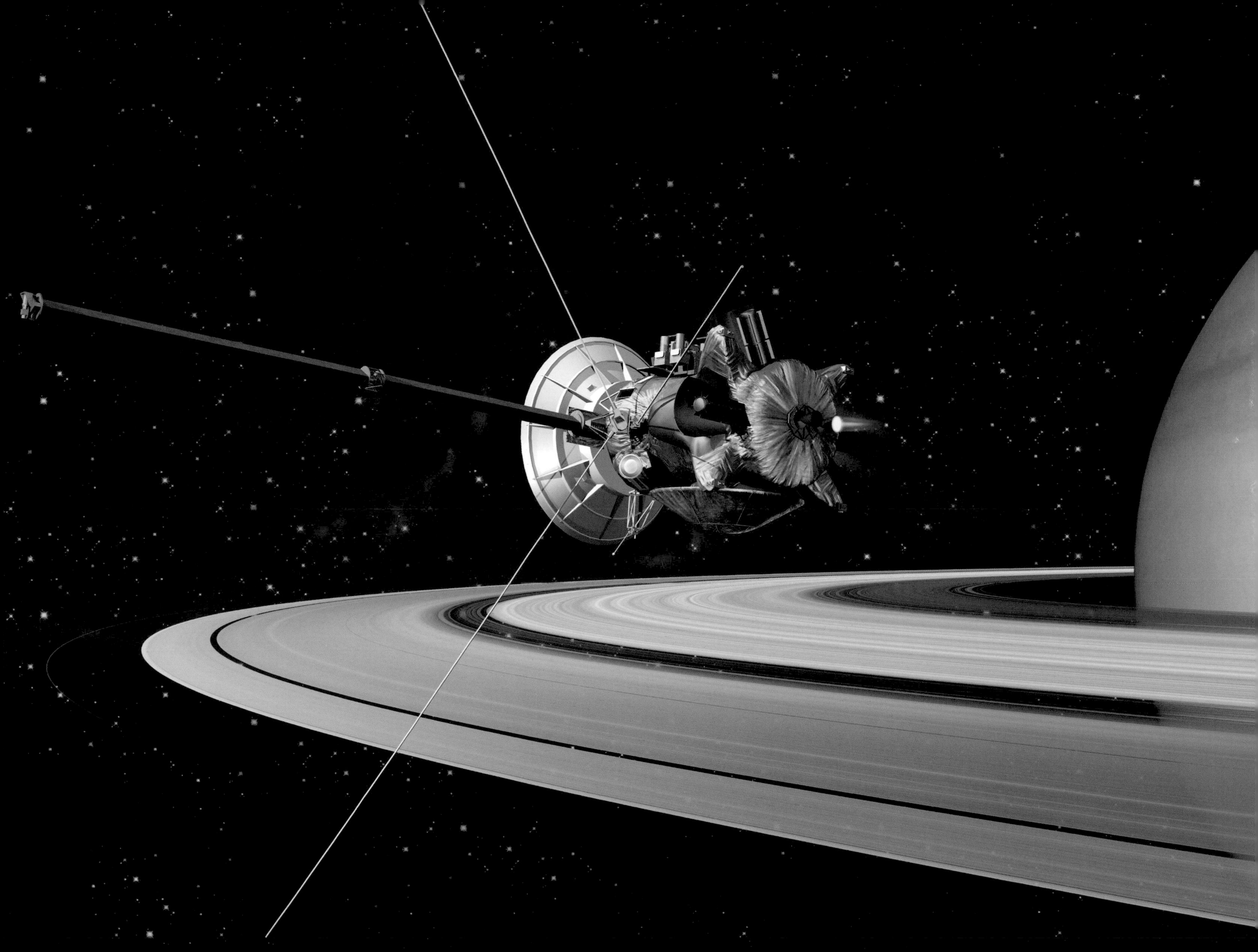

RIGHT & OVERLEAF:

**Seasons on Saturn**

A Saturnian year takes 29 and a half Earth years to complete. The planet is tilted on its axis by 27°, a little more than Earth, so Saturn experiences pronounced seasons. As viewed from Earth, this means that the rings can be seen edge on when Saturn is at an equinox and tipped towards us around a Saturnian solstice. The inclination of *Cassini*'s orbit to Saturn's equator was varied in order to provide the best views of both rings and moons. During its 13 years in orbit, *Cassini* observed both a Saturnian equinox (in 2009, this page) and a northern hemisphere summer solstice (in 2017, overleaf). The angle of sunlight on the rings was very different and seasonal changes were detected in the atmosphere.

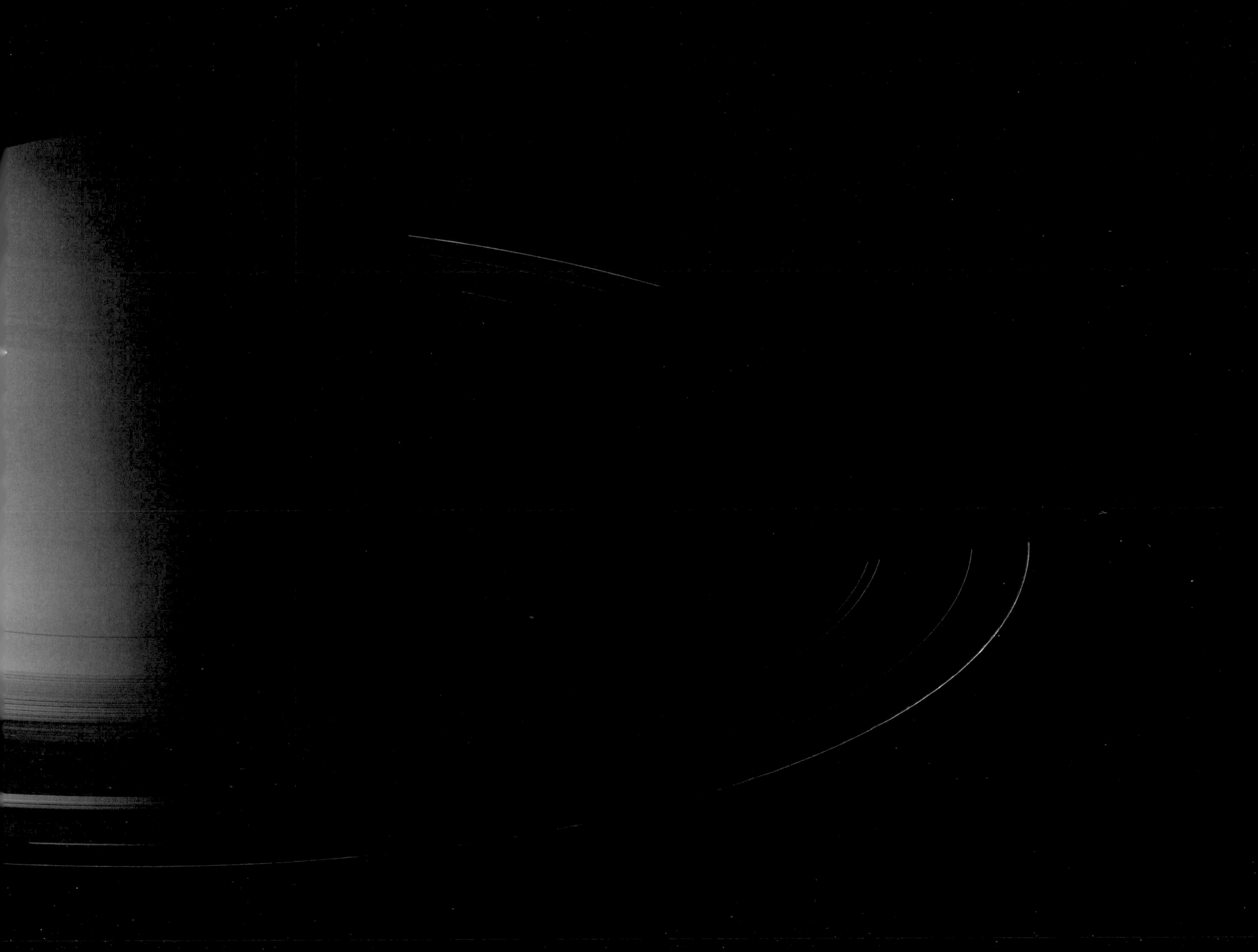

RIGHT:
**Saturn's hexagon**
Discovered by *Voyager 2* and subsequently imaged by *Cassini*, Saturn's hexagon is a persistent cloud pattern at the North Pole, wider than the diameter of Earth. It may comprise a jet stream of 320km/h (200mph) winds.

Laboratory studies suggest that its stable shape is the result of changes in wind speed or direction at different latitudes near the pole. Between 2012 and 2016, the hexagon changed in colour from blue to gold, probably the result of seasonal changes in the atmosphere as the north pole came back into sunlight following the spring equinox in 2009.

ALL:

**Saturn's rings**

Saturn's main rings extend from 7000km to 80,000km (4000–50,000 miles) above its equatorial cloud tops, a width of some 73,000km (45,000 miles). Remarkably their thickness is less than 1km and in places as little as 10m. The rings are composed almost entirely of water ice which makes them extremely reflective and bright.

*Cassini* was able to measure the mass of Saturn's rings as 15 million billion ($1.5 \times 10^{16}$) tonnes, which is less than half the mass of solid ice on Earth spread over a surface area 80 times larger than that of our planet.

Five main rings are recognized (D, C, B, A and F), with fainter G and E rings further out. Undoubtedly one of the most spectacular sights in the Solar System, it is not known how or when they formed. They have been thought to date back to the formation of Saturn billions of years ago. However, data from *Cassini* suggest they may be much younger, within the last 100 million years. Material is being lost from the rings over time and they may disappear within 300 million years from now. If this is correct, we are fortunate to be seeing Saturn at this period in geological time.

The rings could have originated from a former icy moon of Saturn that approached too close to the planet and was ripped apart by tidal forces. The amount of material in the rings is comparable to the mass of the innermost moon, Mimas. Alternatively they could comprise ice left over from the original material out of which Saturn coalesced.

OPPOSITE:

On 15 September 2006, *Cassini* passed into Saturn's shadow. This enabled it to safely turn its cameras sunwards and photograph the Ringed Planet backlit by the eclipsed Sun. This remarkable image is a stitched mosaic of 165 individual pictures taken by *Cassini* over the course of 12 hours. The pale dot between the bright main rings and the fainter G ring is Earth, 1.4 billion km distant.

RIGHT:

*Cassini* acquired this image of Saturn's rings on 3 May 2005, when its orbit carried it behind the rings as seen from Earth. The observed changed of radio signals received from the probe during this occultation enabled the depth of the rings to be measured. The colours represent the ring density as measured by the extent to which radio signals were blocked.

ALL:
**Structure of the rings**
Saturn's rings have amazingly detailed structure. The largest A and B rings (above) are separated by the Cassini Division, a 4800km (3000 miles) wide region of reduced particle density. It is maintained by the moon Mimas, which orbits Saturn once for every two orbits of particles within the Division. This orbital resonance destabilizes the orbits of ice fragments at this distance from Saturn.

Ring particles range in size from micrometres to metres. Radar shows that tidal interactions causes particles in the rings to be sorted by size. We might think of this process as analogous to the way that ocean tides and waves on Earth sort particles on a beach into mud, sand, shingle and cobbles.

The rings display a multitude of subtle colours, captured in these natural colour views from *Cassini*. Hints of pink, grey and beige are thought to result from contamination of their water ice by trace amounts of other materials such as rock or carbon compounds.

LEFT:

**Shepherd moons**

Whilst some gaps in Saturn's rings are created by orbital resonances with moons further out, others are maintained by 'shepherd' or 'herding' moons within the rings themselves. Daphnis is one such shepherd moon, which clears the Keeler Gap within the A Ring.

The tiny moon, just 8km (5 miles) in diameter, creates gravitational ripples on the outer edge of the gap which are visible in this image.

**Saturn's moons**

Saturn has 83 confirmed moons (not counting those embedded in its rings), which is even more than Jupiter. Most are small and irregular in shape, only 13 being larger than 50km (30 miles) in diameter. Fifty-nine of the moons have irregular orbits, suggesting that they are likely to be captured asteroids or other bodies previously in orbit around the Sun. Amongst the 24 with regular orbits, seven are regarded as major moons in that they are large enough to be spherical as a result of their own gravity. All these major moons are tidally locked so that they always present the same face to Saturn, in the same way that our Moon is tidally locked to Earth.

By far the largest is Titan, which represents more than 96 per cent of the combined mass of all Saturn's moons. Titan is the second biggest moon in the Solar System, approaching twice the mass of our own Moon, and the only moon that has a substantial atmosphere.

This mosaic shows the relative sizes of Saturn's nine largest moons in order of increasing distance from Saturn, along with part of the planet itself for scale.

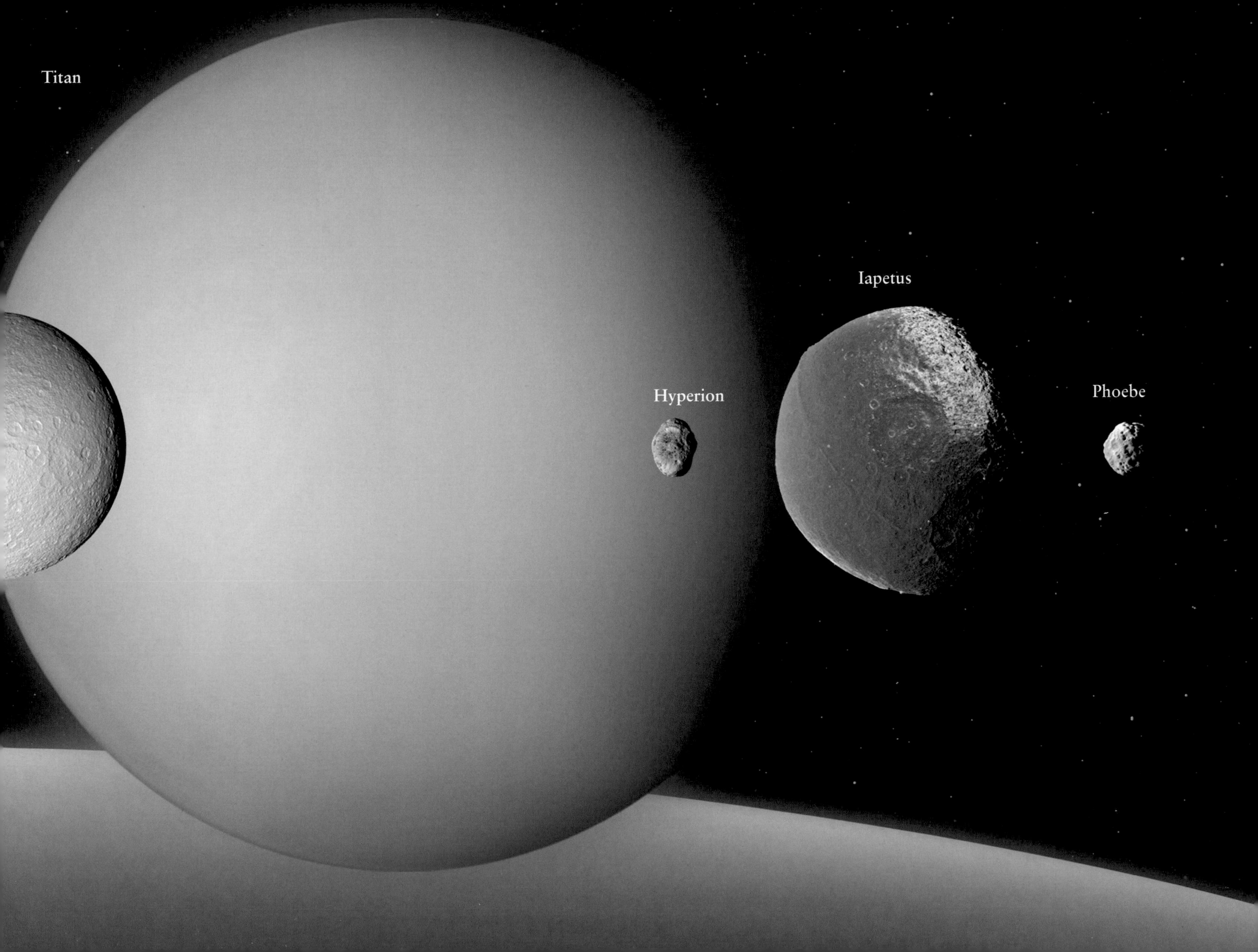
Titan
Iapetus
Hyperion
Phoebe

ABOVE:

**Saturn and Titan**

Titan and Saturn were photographed together, with the rings, in this intriguing image from *Cassini*. Titan takes 16 days to complete each orbit, at a distance of 1.2 million km (750,000 miles) from the planet. Because both the rings and Titan are in equatorial orbits, to an observer on Titan the rings are only ever seen edge on. Their shadow, however, is cast onto the planet so a Titan-based observer could deduce the scale and structure of the rings.

This picture was acquired on 6 May 2012, three years after Saturn's northern hemisphere spring equinox. Consequently the shadow of the rings is cast onto the planet's southern hemisphere, which is taking on a bluish tint as autumn advances. The northern hemisphere, on the other hand, appears yellow as summer approaches, due to atmospheric haze created by increased intensity of ultraviolet sunlight.

ABOVE & OPPOSITE:

**Titan**

With a diameter of 5149km (3200 miles, bigger than Mercury) and a thick atmosphere, Titan seems more like a planet than a moon. It is thought to have a rocky core and a mostly icy mantle and crust. Titan's atmosphere mostly comprises nitrogen (97 per cent, compared to 78 per cent on Earth) and a surface pressure 45 per cent higher than that of Earth. Its surface temperature, however, is a chilly -179°C (-290°F). On Titan, water is frozen so hard it behaves as rock. The surface is shrouded by a permanent layer of haze comprising methane and other simple carbon-based compounds, which filter 90 per cent of incoming sunlight.

Pictured above is a composite infrared image of Titan acquired by *Cassini* during a fly-by on 13 November 2015. Infrared enables us to peer through atmospheric haze and reveals some surface detail.

Pictured opposite are six infrared images of the surface of Titan using data acquired by *Cassini* in multiple fly-bys.

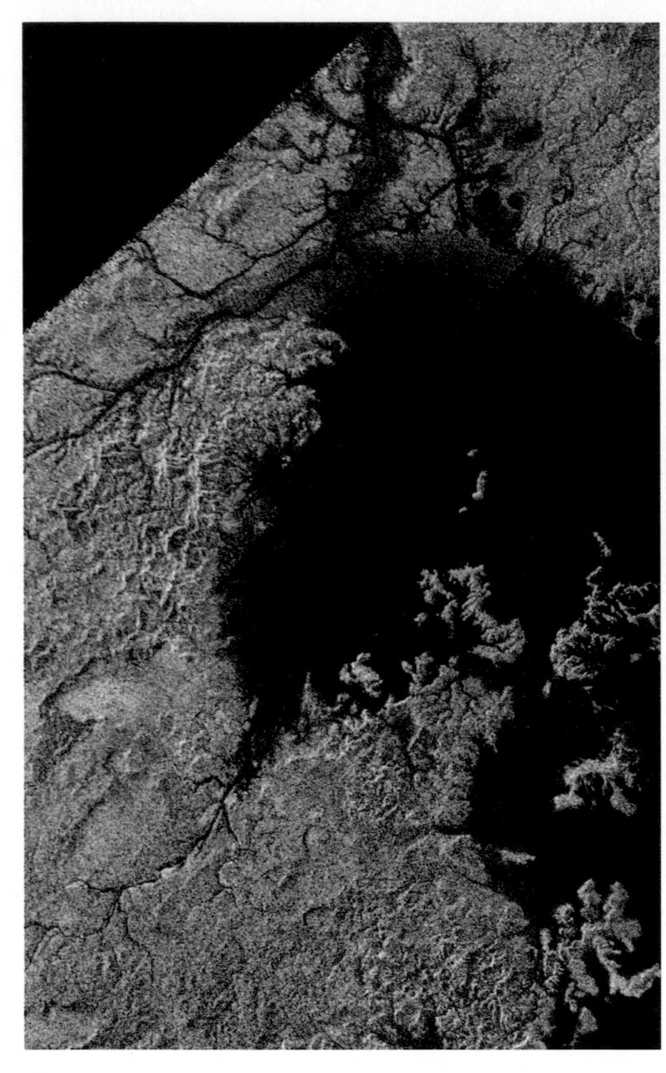

ABOVE:
**Seas on Titan**
One of the most remarkable discoveries made by *Cassini* is that Titan has seas and lakes of liquid hydrocarbons, predominantly methane and ethane, on its surface. This radar image shows Ligeia Mare in Titan's north polar region, the second-largest sea discovered. These seas and lakes are the only known stable bodies of surface liquid known to exist outside Earth. It seems that liquid hydrocarbons fall as rain and flow as rivers (some visible in this image) on the surface of Titan. These processes are closely analogous to Earth's hydrological (water) cycle but occur at temperatures that are 200°C (392°F) colder.

LEFT:
**Surface of Titan**
The surface of Titan photographed by *Huygens*. The sand, rocks and pebbles are made of water ice, rounded by flowing hydrocarbons.

RIGHT:
**Landing on Titan**
The *Cassini* spacecraft carried with it a probe named *Huygens* whose mission was to parachute through Titan's atmosphere and land on the surface. This audacious enterprise was successfully completed on 14 January 2015, the most distant landing ever achieved by a spacecraft and the first on any moon other than our own. The probe was battery-powered and designed to operate for just three hours after being 'woken up' by an on-board timer just before atmospheric entry. It transmitted photographs and data from its descent and the surface to *Cassini*, which relayed them to Earth.

*Huygens* revealed that Titan's surface is geologically young, with few craters. There are hills and rocks made of water ice, exhibiting abundant evidence of fluvial erosion. Liquid hydrocarbons flow across the surface in much the same way that water does on Earth. Because of Titan's distance from the Sun and thick atmospheric haze, light levels on the ground are only one-thousandth of those on a sunlit day on Earth.

The presence of seas, lakes, rivers and rain on Titan have led to speculation that conditions could be suitable for some form of life. With hydrocarbons rather than water as the solvent, the chemistry of any such life would be radically different to that on Earth. Whether life in any form exists is a question for future exploration. The next such mission is called *Dragonfly*, scheduled to launch in 2027, arrive at Titan in 2034 and fly a nuclear-powered drone in the atmosphere. Other concepts being developed are for a hot-air balloon to drift through Titan's atmosphere and a lander to float on the surface of a hydrocarbon lake.

Pictured is an artist's impression of *Huygens* parachuting to the surface of Titan.

OVERLEAF:
A view of the surface of Titan imaged by *Huygens* as it descended. The landing site appears to be a dry lake bed.

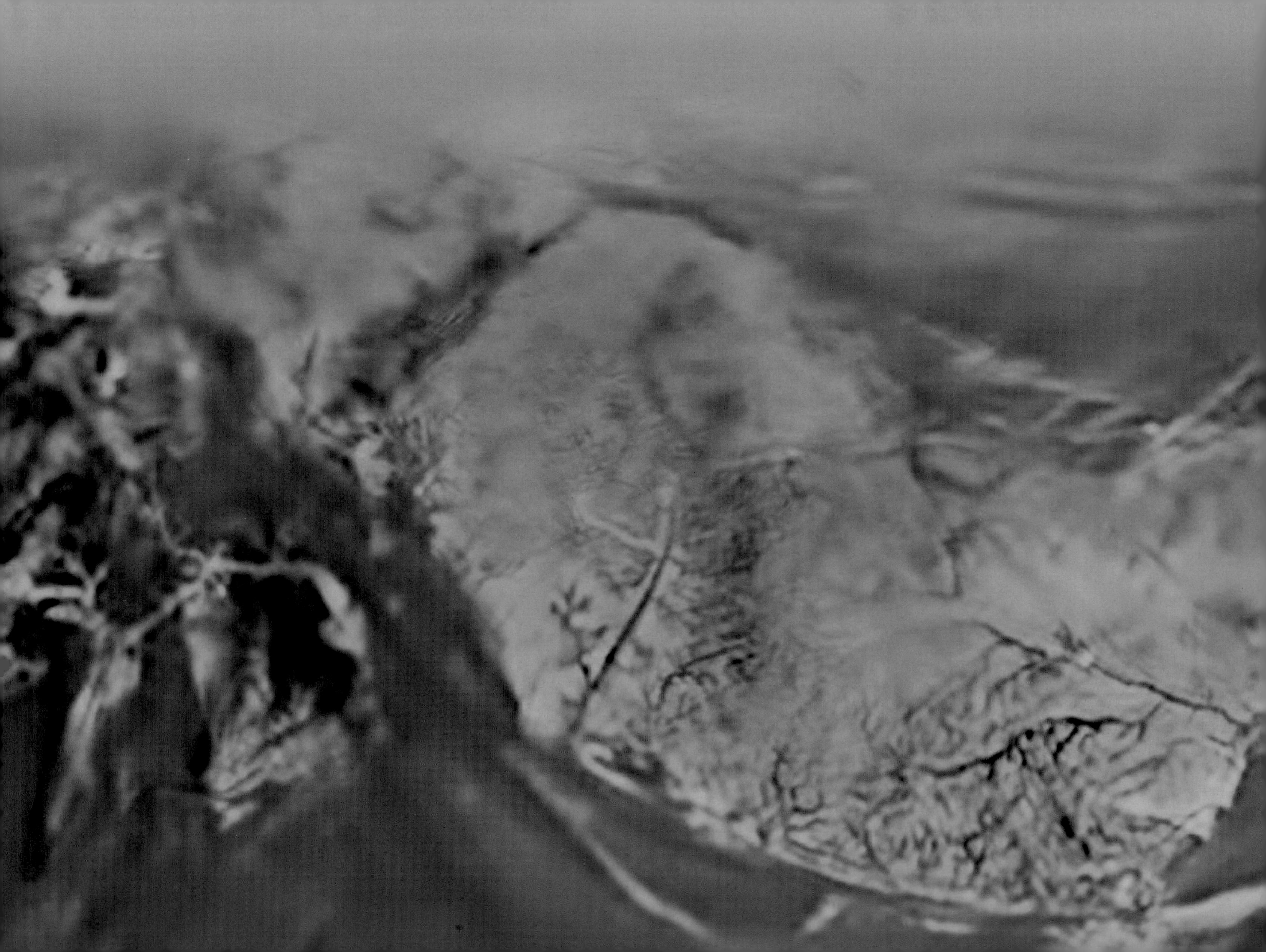

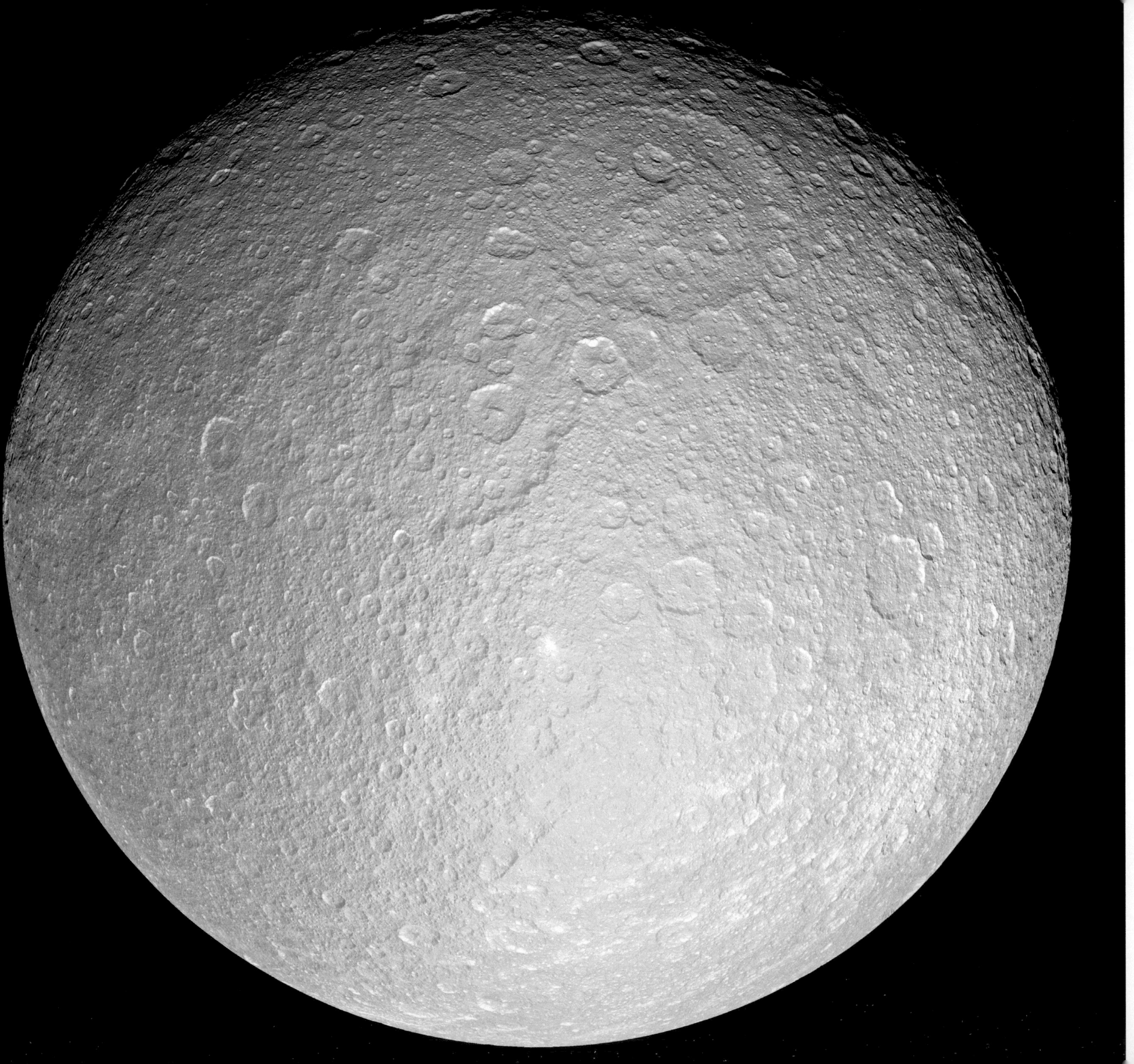

LEFT:
**Rhea**
Saturn's second-largest moon, Rhea, is 1528km (950 miles) in diameter, 527,000km (330,000 miles) from the planet and orbits in four and a half days. It is predominantly made of ice with a small rocky core. The surface is heavily cratered but the distribution of larger craters suggests that extensive resurfacing occurred at some stage in its history.

ABOVE:
**Iapetus**
Iapetus is a most curious world; half is as light as snow, whilst its leading hemisphere is black. The explanation appears to be that another moon, Phoebe, which orbits in the opposite direction, has deposited a ring of dust and ice in its orbit. Iapetus has collided with these particles, which have darkened its forward-facing hemisphere. The darker hemisphere absorbs more sunlight and warms up, leading to ice sublimating and being redeposited on the trailing hemisphere. Iapetus is Saturn's third-largest moon, 1436km (900 miles) in diameter and orbits 3.5 million km (2.2 million miles) from Saturn in 79 days.

**Phoebe**

Phoebe is a much smaller, irregular moon, 220km (137 miles) in diameter in a retrograde orbit. Unlike Saturn's other moons it is dark in colour. Phoebe is believed to be a captured centaur, originating from the outer Solar System beyond Neptune. It is much further out than Iapetus, at a distance of 13 million km (8 million miles), with an orbital period of 547 days. Dust ejected from Phoebe's surface by micrometeoroid impacts spirals inwards towards Saturn where it meets Iapetus.

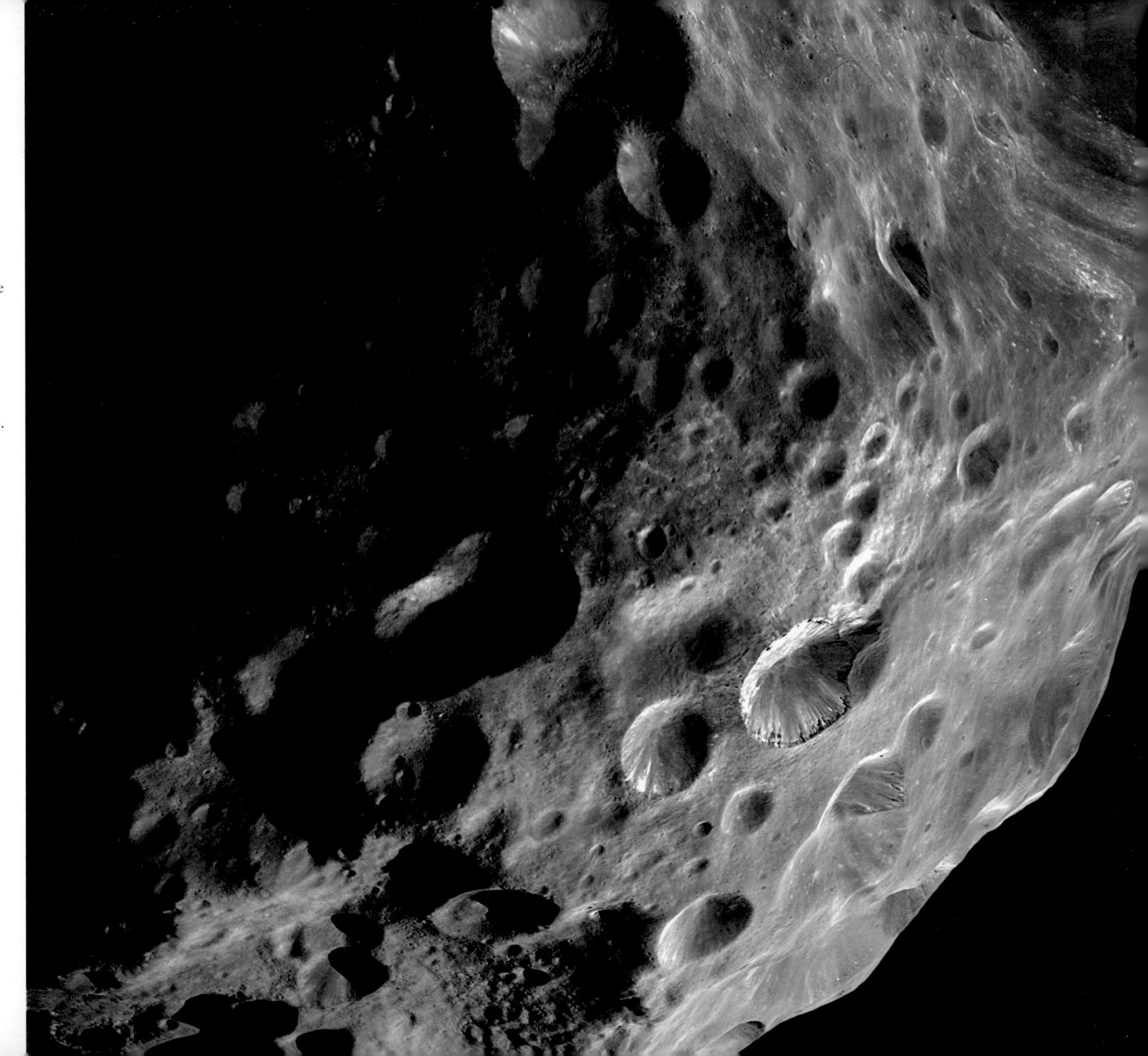

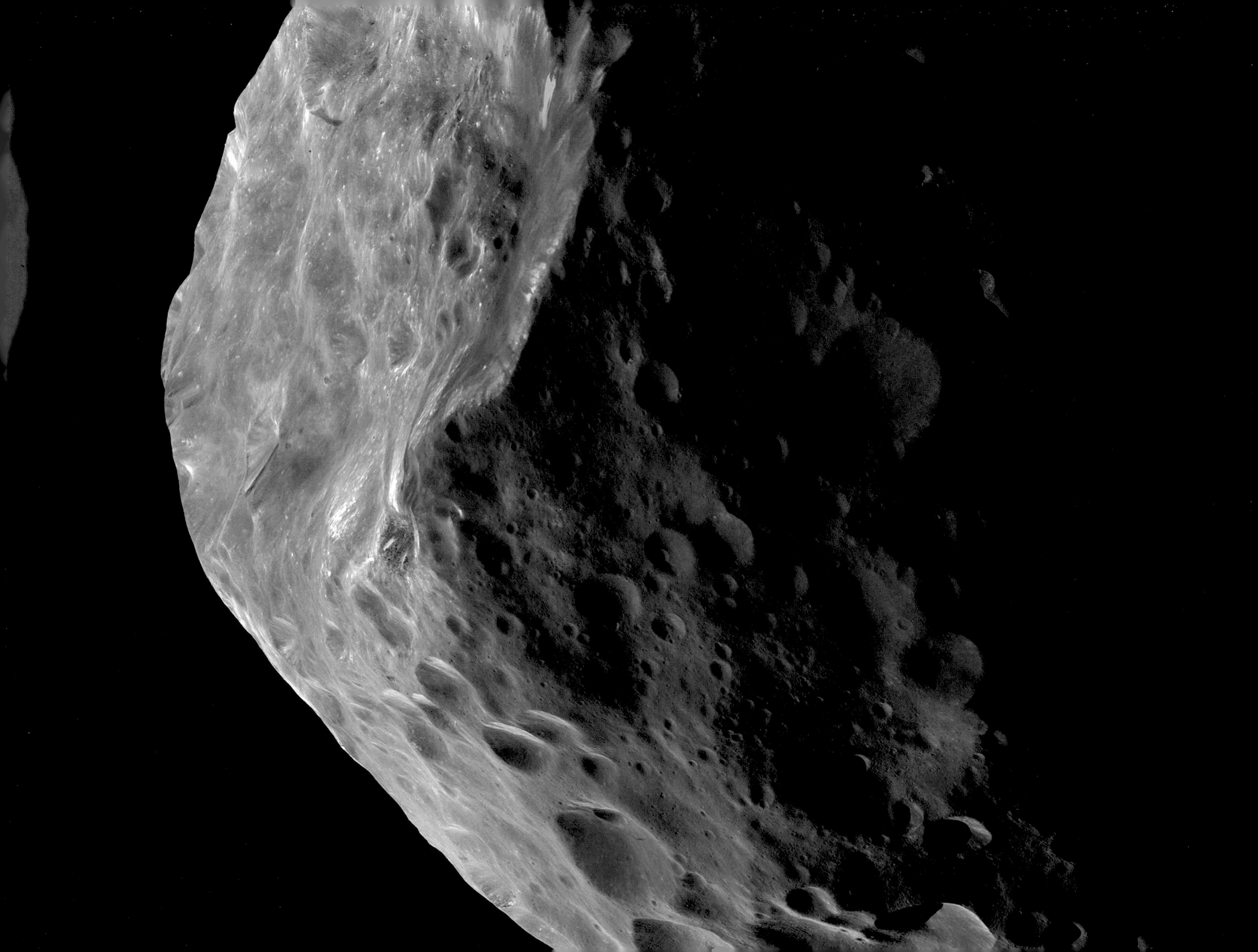

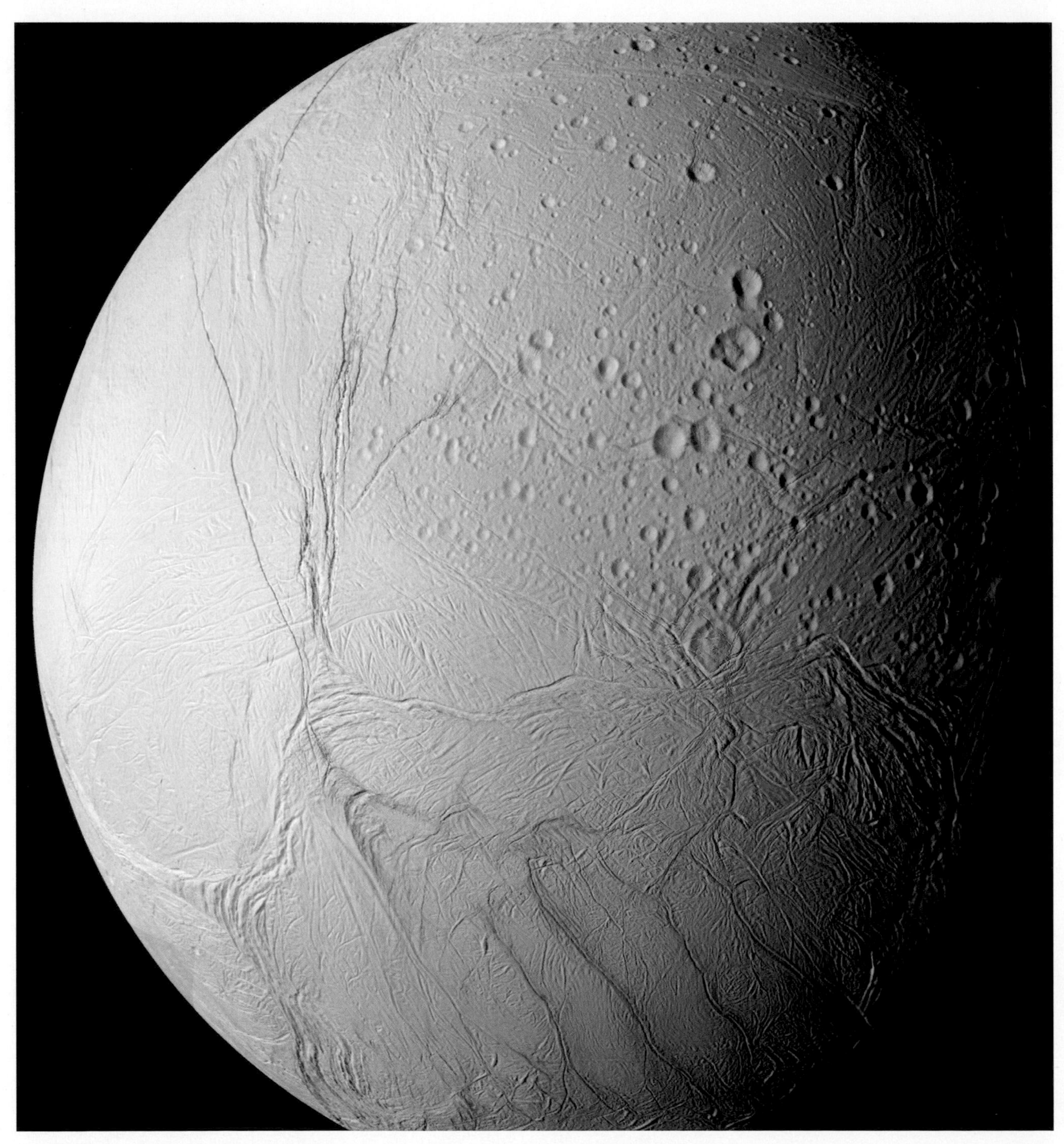

### Enceladus

One of the closest moons to Saturn, Enceladus is covered in pure water ice which makes it one of the most reflective objects in the Solar System. There are relatively few craters, indicating a geologically young surface. Near the south pole are a series of dark markings referred to as 'tiger stripes'. *Cassini* discovered icy particles and water being jetted into space from these stripes, superficially similar to geysers on Earth. This indicates that there is liquid water beneath the icy crust, which is believed to comprise an ocean up to 30km (18 miles) deep beneath a similar depth of ice.

The source of heat liquefying this ocean and driving the geysers (more accurately referred to as cryovolcanoes) appears to be tidal resonance with the larger moon Dione, which orbits exactly once for very two orbits of Enceladus. With a diameter of 500km (310 miles), Enceladus is the smallest known geologically active body in the Solar System. It orbits Saturn in 1.3 days at a distance of 240,000km (150,000 miles).

In 2015, *Cassini* flew through one of the water vapour plumes and detected molecular hydrogen, suggesting the presence of hydrothermal vents on the subsurface ocean floor. This in turn has led to the identification of Enceladus as a possible habitat for water-based life. Further missions are being developed to investigate this intriguing possibility.

Pictured left is Enceladus, showing an older cratered region and younger smooth ice.

Pictured below are geysers jetting water and ice from the south polar region of Enceladus.

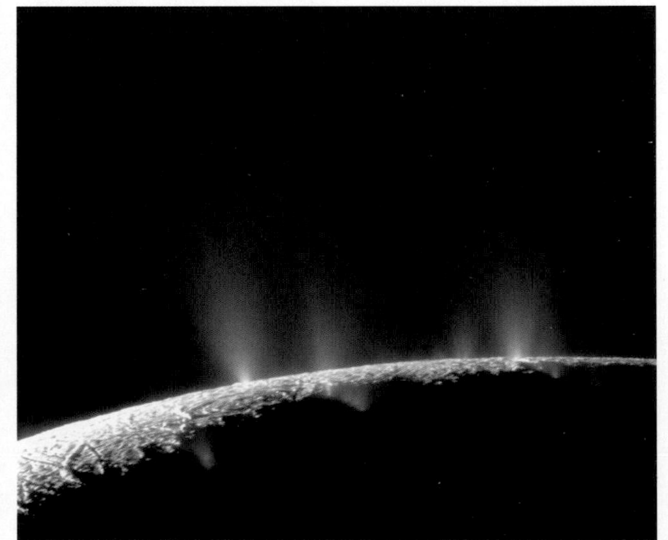

ABOVE:
**Mimas**
The smallest and innermost major moon, Mimas orbits Saturn in just under one day at a distance of 186,000km (116,000 miles). In contrast to Enceladus, it is geologically inactive. Its diameter of 396km (246 miles) makes Mimas the smallest known object in the Solar System to be approximately spherical as a result of its own gravity. The most notable surface feature is a huge crater, 130km (80 miles) across, named Herschel after the astronomer who first discovered Mimas.

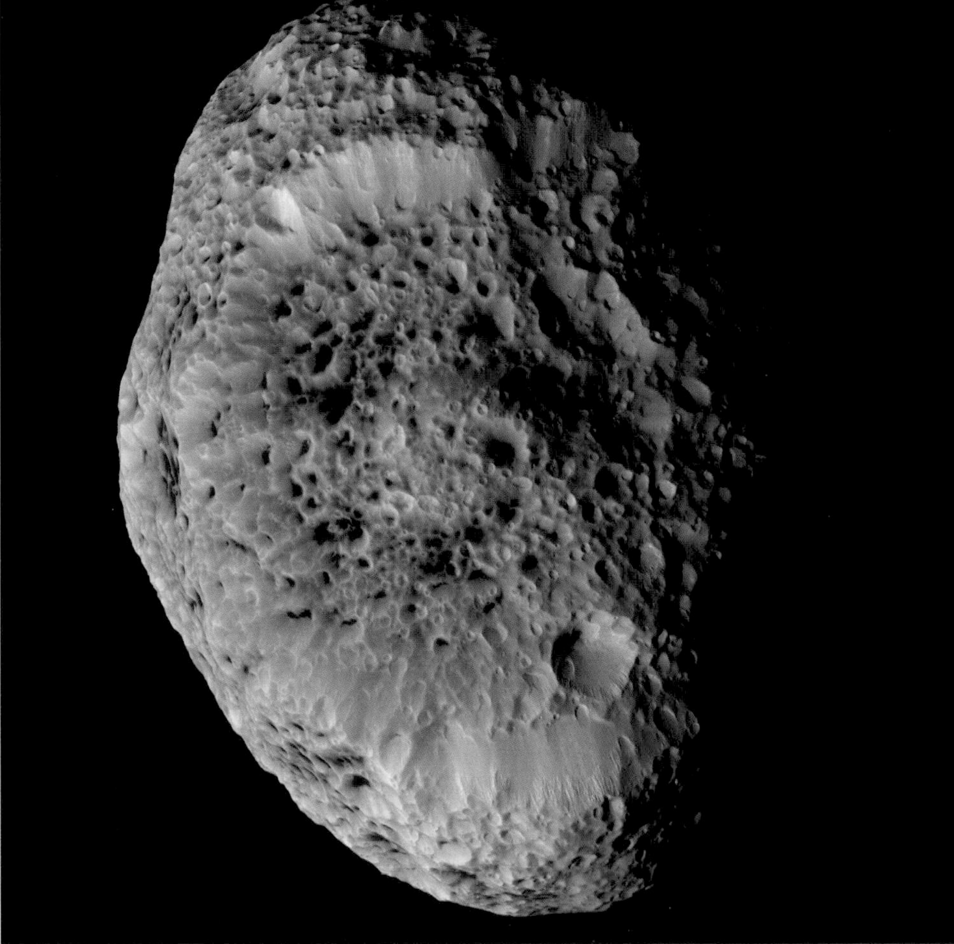

ABOVE:
**Hyperion**
Hyperion is the largest irregularly-shaped moon in the Solar System, 360km (220 miles) across its longest axis. Its surface resembles a sponge. With an overall density just half that of water ice, it must be highly porous internally. In contrast to other Saturnian moons, the surface is tan-coloured, perhaps as a result of material from other moons being deposited on its surface. At a distance of 1.5 million km (900,000 miles) from Saturn, Hyperion is locked in tidal resonance with Titan, making exactly three orbits for every four of its larger neighbour. As a result of this tidal resonance Hyperion is notable for being the only planetary moon known to exhibit chaotic rotation; it has no well-defined poles or equator.

# Uranus & Neptune

Uranus and Neptune are ice giants, made up predominantly of elements heavier than hydrogen and helium. Compounds such as water, ammonia and methane exist as solids in the cold outer reaches of the Solar System and comprise the bulk of these planets.

The two ice giants are similar in size and mass. Uranus has the larger diameter (51,118km/31,763 miles, four times that of Earth) whilst Neptune is slightly more massive (17 times the mass of Earth). Neither were known in antiquity. Uranus was discovered in 1781 by William Herschel and Neptune in 1846 by Johann Galle as a result of predictions made by Urbain Le Verrier and John Couch Adams. Uranus orbits the Sun in 84 years and Neptune in 165 years, so neither planet has yet completed three orbits since its discovery.

Uranus has 27 known moons and a system of 13 rings. Perhaps its most curious feature is its extreme axial tilt of 98°. This means that its poles are almost aligned with its orbit, creating seasons unlike those of any other planet. Neptune has fainter rings and 14 known moons, of which by far the most significant is Triton. One of the largest and most bizarre natural satellites in the Solar System, Triton is thought to have been a dwarf planet gravitationally captured by Neptune.

Only one spacecraft has journeyed to the ice giants. *Voyager 2* flew past Uranus in 1986 and then Neptune in 1989. This 'Grand Tour' of the outer Solar System was possible thanks to an alignment of all four giant planets that occurs only once every 175 years. Jupiter's gravity was used to boost the probe onto a faster trajectory, cutting several years off the travel time to Uranus and Neptune.

OPPOSITE:

**Uranus**

The green colour tinge of Uranus is discernible through Earth-based telescopes. *Voyager 2* revealed a rather bland pale green disc. Like the gas giants, its atmosphere comprises mainly hydrogen and helium. The green tinge results from methane, the third most abundant component. Other hydrocarbons are formed by the action of sunlight on methane, creating an atmospheric haze that contributes to the lack of visible features. Above its visible atmosphere, Uranus has an extended corona of hydrogen reaching 50,000km (30,000 miles) into space, a unique feature amongst the planets. Uranus has no solid surface but the interior comprises mostly water, ammonia and methane ices, together with rock. At 2.9 billion km (1.8 billion miles) from the Sun (19 times further than Earth), it is the coldest planet, with a minimum atmospheric temperature of -224°C (-371°F). The planet's internal rotation rate is 17 hours 14 minutes.

OPPOSITE:

**Seasons on Uranus**

In contrast to all other planets, Uranus' axis of rotation is tilted sideways, at an angle of 98° to its orbit. This is thought to result from a collision with another, perhaps Earth-sized, planet early in its history. Uranus therefore experiences the most extreme seasons of any planet. At solstice, one pole is pointed towards the Sun and the opposite hemisphere is in darkness. Since a Uranian year lasts 84 Earth years, winter would be a grim time on the planet.

*Voyager 2* measured a magnetic field and found it to be highly asymmetric. The field is tilted at 59° from its axis of rotation and generated not in the centre of the planet but at a relatively shallow depth below the South Pole. It may be that an ocean of water and ammonia outside the core is the source of the magnetic field. Aurorae have been seen on Uranus but not around the poles.

This image of crescent Uranus, a view never seen from Earth, was captured by *Voyager 2* as it departed the planet after approaching within 82,000km (51,000 miles) of the cloud tops on 24 January 1986.

RIGHT:

**Rings of Uranus**

Like other giant planets, Uranus has rings. They were first discovered with Earth-based telescopes and investigated in detail by *Voyager 2*. Thirteen rings have now been identified, extending up to 50,000km (30,000 miles) from the cloud tops. Unlike those of Saturn, Uranian rings are dark, reflecting only five per cent of the light that falls on them. They are thought to comprise water ice mixed with darker material such as hydrocarbons. The rings likely originate from break up of a small moon due to collision or gravitational disruption.

Epsilon is the brightest ring, clearly visible in this infrared, false-colour image from *Hubble Space Telescope*. It is thin (less than 100km/60 miles wide) and narrow (possibly less than 150m/500ft). Particles in the ring are confined by two shepherd moons, Cordelia and Ophelia, orbiting just within and beyond the ring.

## Moons of Uranus

Uranus has 27 known moons, of which five are bigger than 450km (280 miles) in diameter and spherical in shape. All take their names from characters in plays by William Shakespeare and a poem by Alexander Pope.

The five major moons (Miranda, Arial, Umbriel, Titania and Oberon) orbit in Uranus' equatorial plane and share its axial tilt. This indicates they formed from an accretion disc after the giant impact that knocked the planet sideways. They are composed of rock and ice. Titania is the largest, with a diameter of 1578km (980 miles), and is about half the mass of our Moon.

Within the orbit of Miranda are 13 small inner moons such as Puck. They are closely linked to the rings and likely made of similar material. Due to gravitational interactions, their orbits are unstable and the moons are likely to collide from time to time.

Nine irregular moons lie outside the orbit of Oberon. All are small (less than 200km/120 miles in diameter), have eccentric orbits and eight of them orbit in the opposite direction to the rotation of Uranus. These characteristics indicate that they were captured by Uranus after the planet was formed.

This mosaic (pictured right) shows Uranus's five largest moons to scale (though not their relative positions), along with part of Uranus itself.

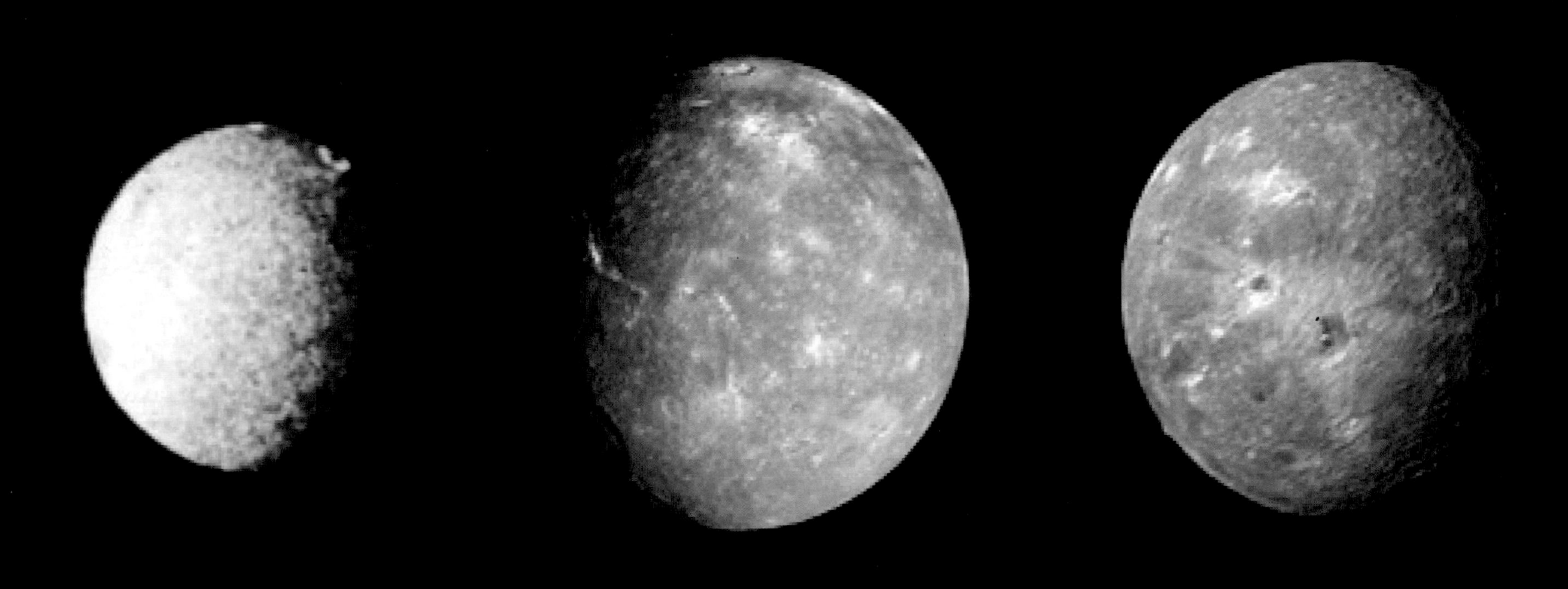

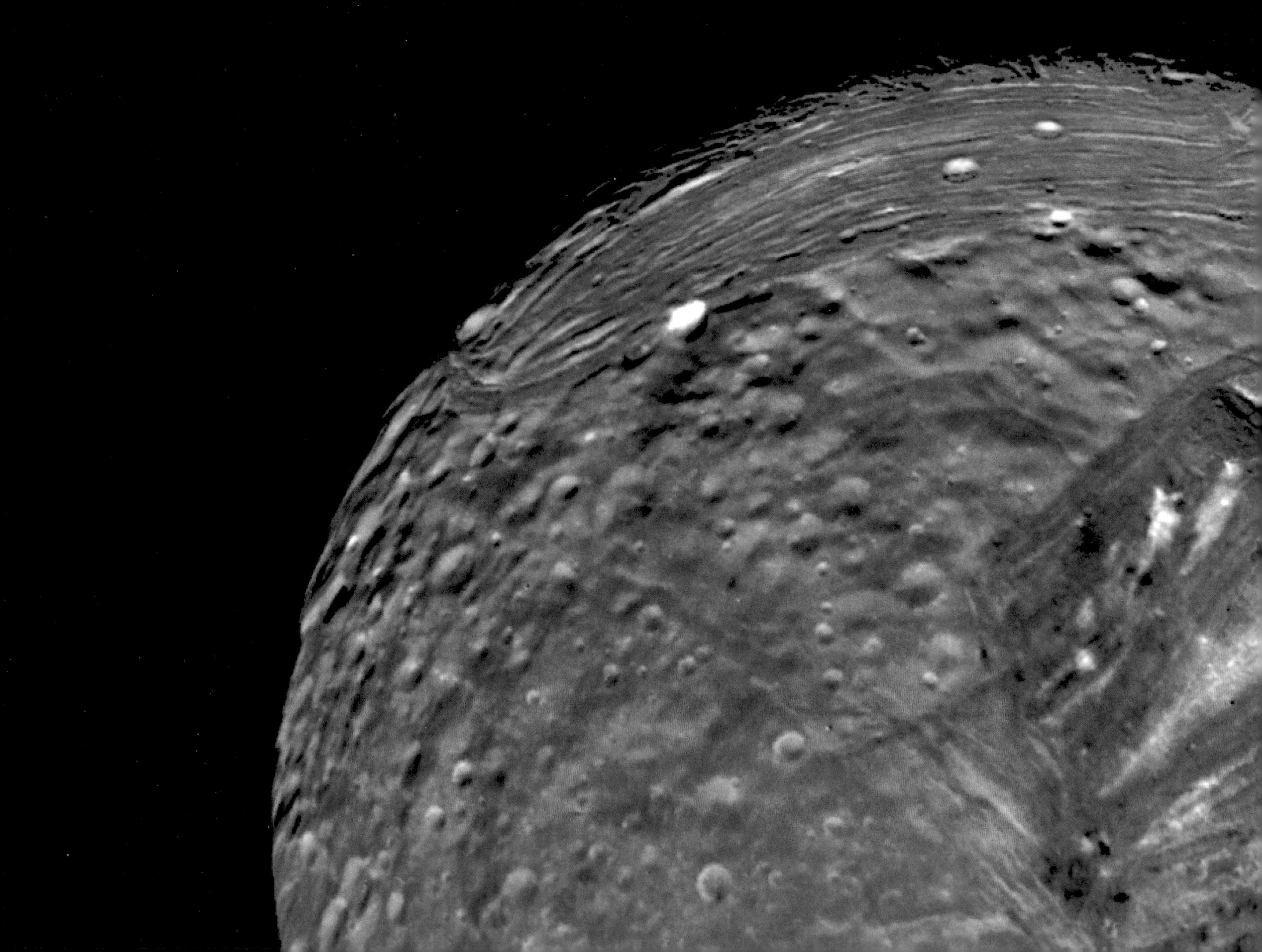

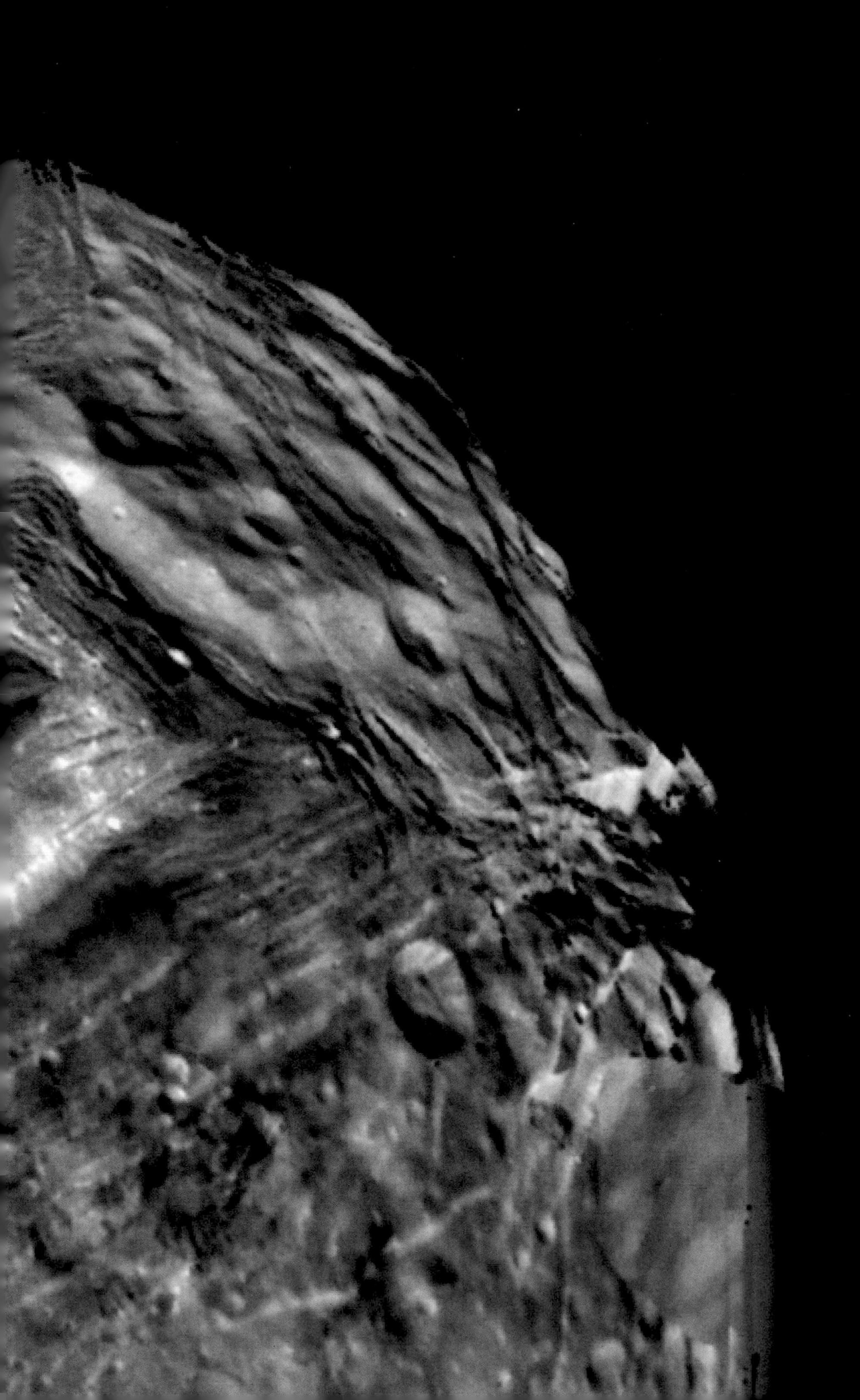

## Miranda

The most curious of Uranus's moons is Miranda, which has some of the most extreme topography in the Solar System. Just 470km (290 miles) in diameter, it orbits the planet in 34 hours, 129,000km (80,000 miles) above its cloud tops. *Voyager 2* photographed the southern hemisphere, which was sunlit at the time of its fly-by, passing closer than it did to any other Uranian moon.

Miranda's surface has deeply broken terrain and is crossed by huge canyons, hundreds of kilometres in length and tens of kilometres wide. Veronica Rupes is the Solar System's highest known cliff, with a height of 20km (12 miles). Three giant grooved structures, 200km (124 miles) wide and 20km (12 miles) deep, called coronae are unique to Miranda.

Initially it was a mystery how such dramatic features could form on such a small moon. It almost looks as if Miranda may have been shattered by a giant impact and then reassembled itself as a jumble of rocks. However, the most likely explanation is that in the past Miranda was subject to extreme tidal heating, causing its ice to melt and upwell from the interior, accompanied by cracking and subsidence elsewhere. The source of this heat could have been an orbital resonance with Arial and/or Umbriel, which has now been lost allowing Miranda to cool and solidify.

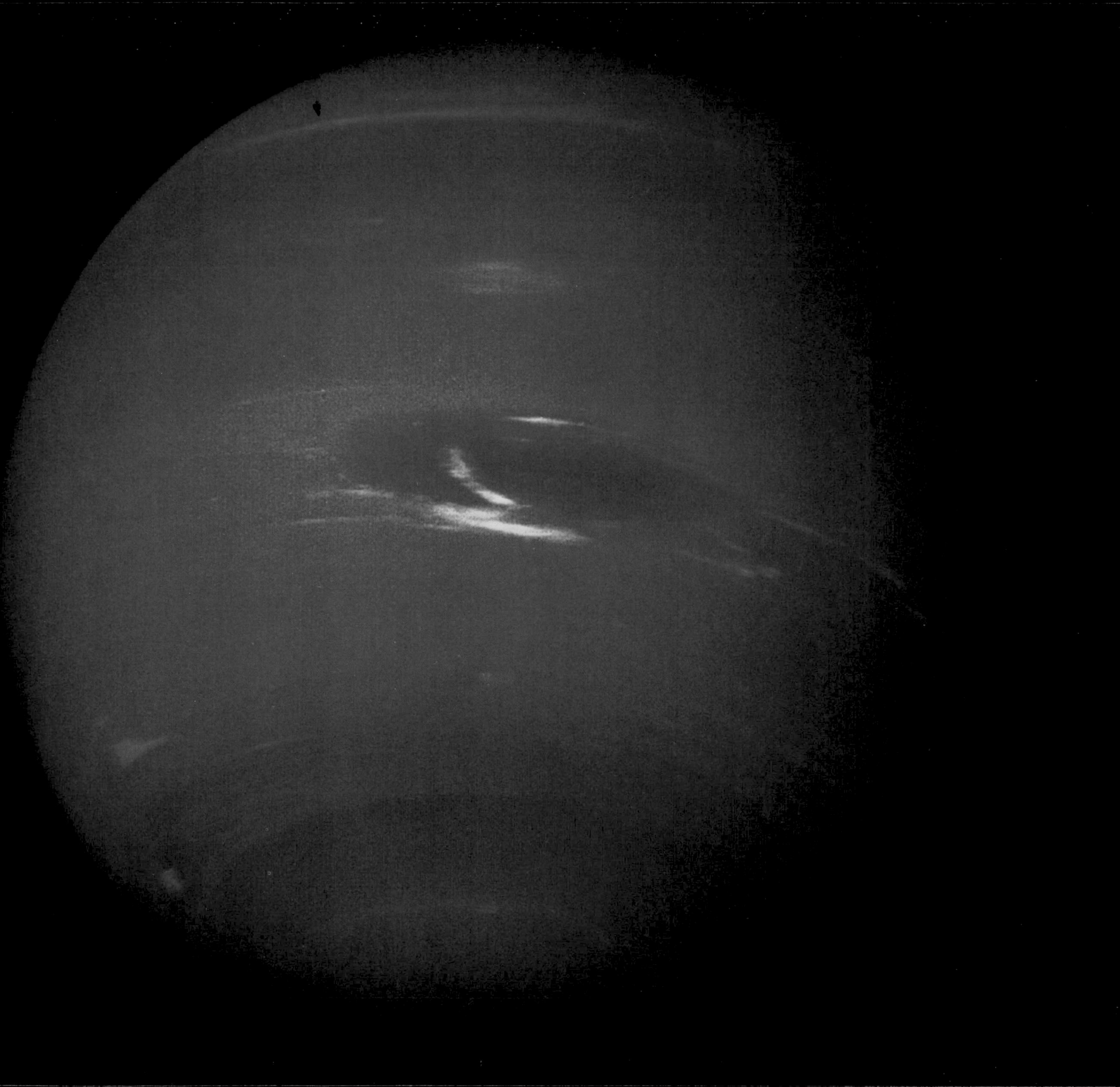

LEFT:

**Neptune**

There are two predominantly blue planets in the Solar System: our own and Neptune. However, the similarities end there as Neptune is an ice giant, 30 times further from the Sun than Earth. At four and a half billion kilometres, this most distant planet takes 165 years to complete a single orbit. Its outer atmosphere is a frigid -218°C (-360°F), at which temperature methane freezes.

The size and composition of Neptune are similar to Uranus, having an interior of water, ammonia and methane ices, together with rock. Neptune's axial tilt is a more modest 28° (similar to Saturn) but its magnetic field is nonetheless inclined at 47° to its poles. This suggests that a tilted magnetic field arises from the internal structure of ice giants rather than their axis of rotation. Neptune's rings are fragmented and fainter than those of Uranus.

Neptune is not a solid body and different parts of the planet rotate at different speeds. The equatorial zone takes 18 hours, the polar regions 12 hours and the magnetic field 16.1 hours, which is the best indication of the core rotation rate. These are the greatest variations in spin within any planet.

LEFT:

**Neptune's atmosphere**

The upper atmosphere is 80 per cent hydrogen and 19 per cent helium, with trace gases such as methane giving the planet its colour. Neptune's atmosphere shows considerably more structure and meteorological activity than Uranus. Greater internally-generated heat drives its weather patterns, giving rise to the Solar System's strongest winds of up to 2100km/h (1300mph).

*Voyager 2* observed a Great Dark Spot, an anticyclonic storm similar to Jupiter's Great Red Spot. It was no longer visible in images from *Hubble Space Telescope* five year later. This and other similar features are thought to be vortexes forming holes in the upper cloud decks to reveal patches of darker troposphere. Conversely, high altitude clouds are bright in colour and cast shadows on the blue cloud deck below.

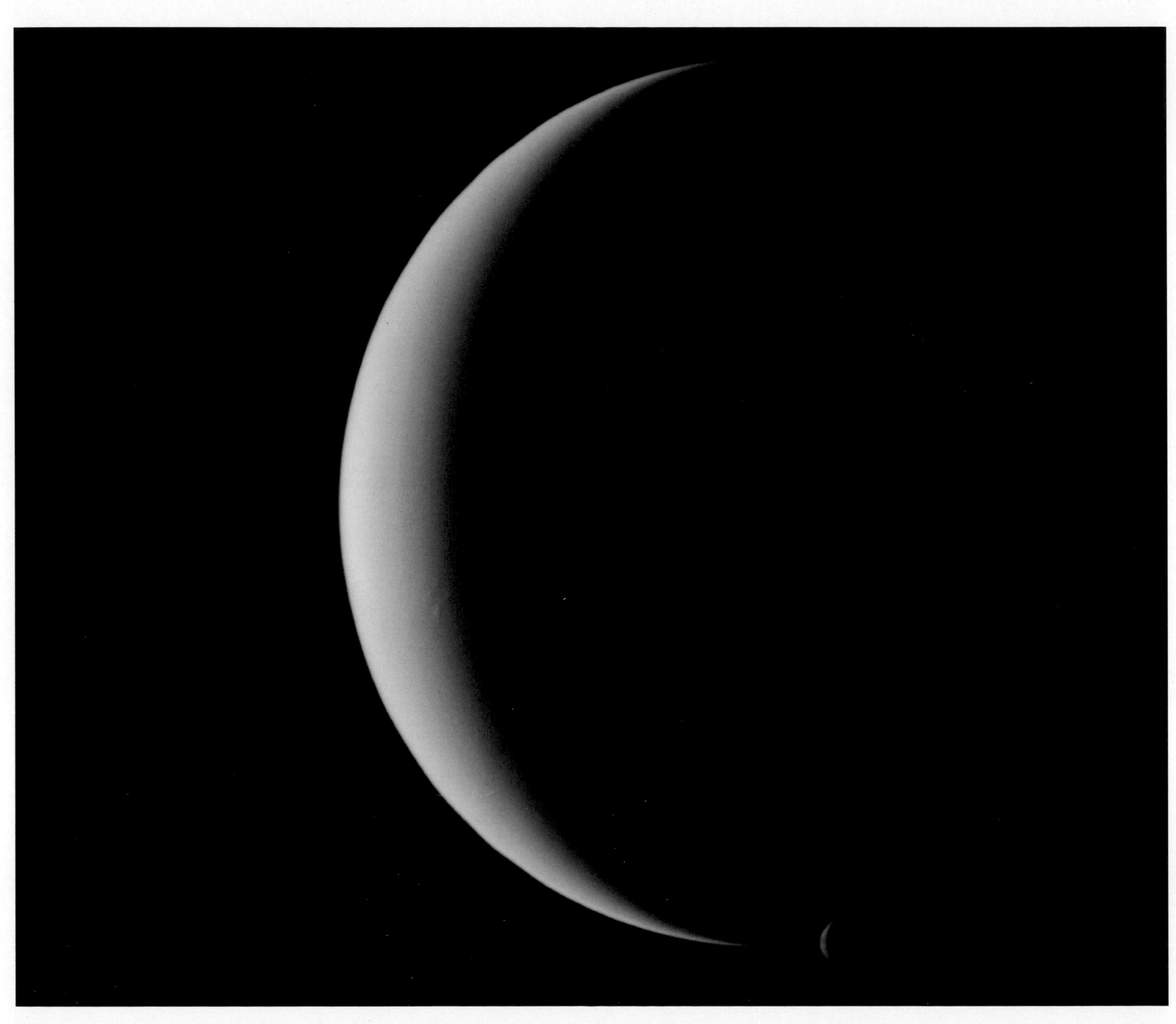

LEFT:
**Neptune's moons**
Neptune has 14 known moons, 13 of which are less than 450km (280 miles) in diameter and irregular in shape. By far the largest is Triton, seen in this image as a crescent along with Neptune. Of the other moons, seven orbit close to the planet and are associated with Neptune's rings. The remainder have irregular orbits (being in the opposite direction to Neptune's rotation, or highly elliptical, or highly inclined to Neptune's equator), indicating they were gravitationally captured. Since Neptune was the mythological god of the sea, all the moons are named after Greek or Roman sea deities.

OPPOSITE:
**Triton**
Triton represents over 99.5 per cent of the mass of all Neptune's moons. With a diameter of 2710km (1680 miles) it is the seventh largest satellite in the Solar System and three-quarters the diameter of Earth's Moon. *Voyager 2*'s trajectory was optimized to provide good images of this intriguing object. Triton is an icy world with a surface temperature of -235°C (-391°F). Its surface is covered in a thin sheet of frozen nitrogen, whilst the crust and mantle comprise water ice overlying a substantial rocky core.

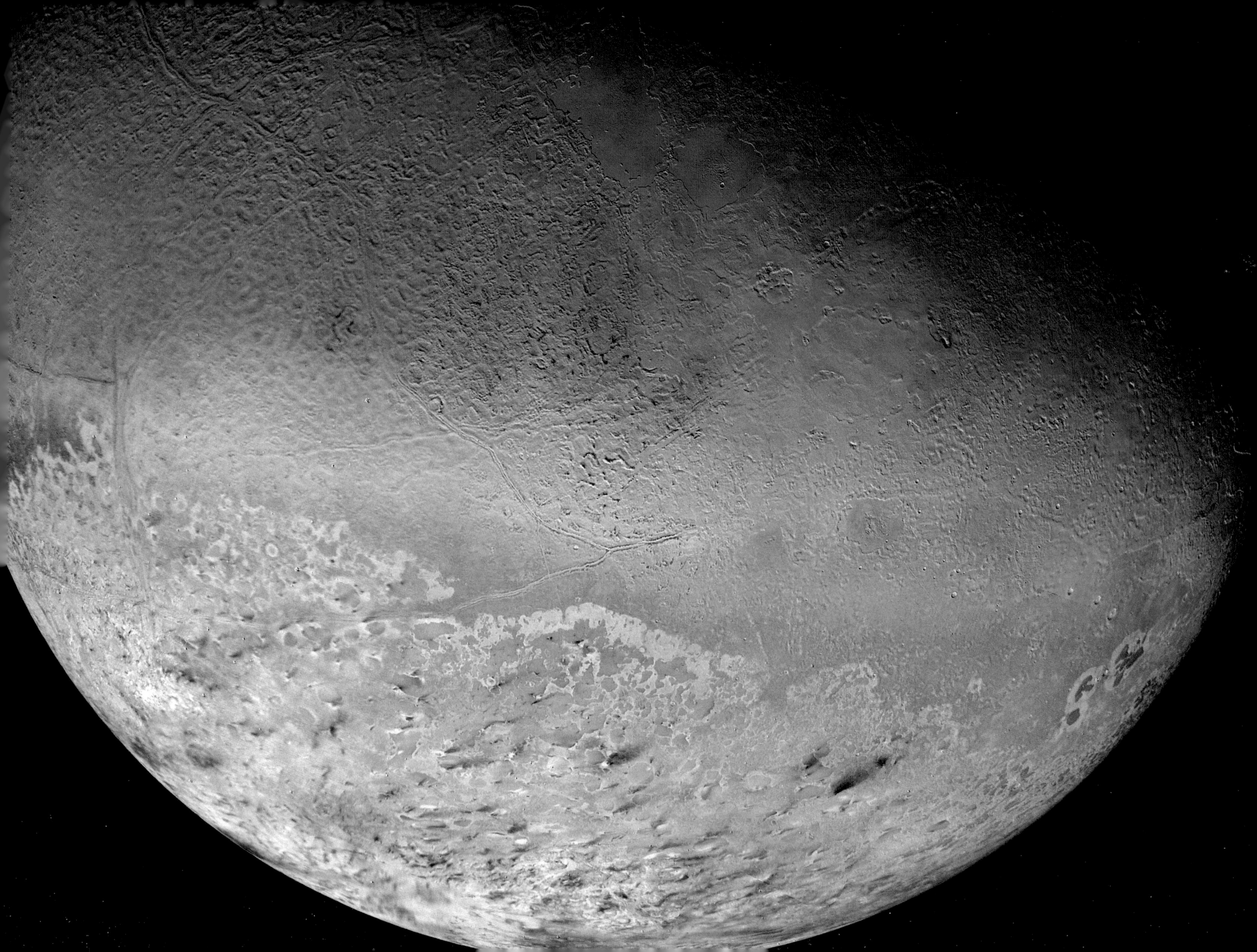

RIGHT:

### The surface of Triton

Triton has a complex surface with few craters, indicating resurfacing by internal geological processes. Its western hemisphere is marked by fissures and depressions, known as 'cantaloupe terrain' as it looks like the skin of a melon. *Voyager 2* found active geysers erupting nitrogen, contributing to a tenuous atmosphere of gaseous nitrogen. This makes Triton one of four moons in the Solar System on which, like on Earth, active eruptions have been observed. The heat that powers these geysers is believed to come from radioactivity in the moon's rocky core and tides generated by the obliquity of Triton's orbit to Neptune's rotation.

BELOW:

### The origin of Triton

Uniquely amongst the Solar System's major moons, Triton has a retrograde orbit (opposite to the rotation of Neptune). This means it cannot have formed from the same nebula as the planet but must have been captured from elsewhere. Triton appears to have originated as a Kuiper Belt object orbiting the Sun, which would have been classed as a dwarf planet in its own right. In fact, Triton is similar in size and composition to Pluto (which is slightly smaller), supporting this theory. To be captured by Neptune, Triton could have collided with an existing moon, causing it to slow down and enter orbit. More likely it was one of a binary pair of Kuiper Belt objects, the other of which was flung away as a result of the encounter, leaving Triton in orbit. The capture of such a large moon would have created gravitational havoc amongst the existing moons of Neptune, which is probably why the planet now has no other major moons.

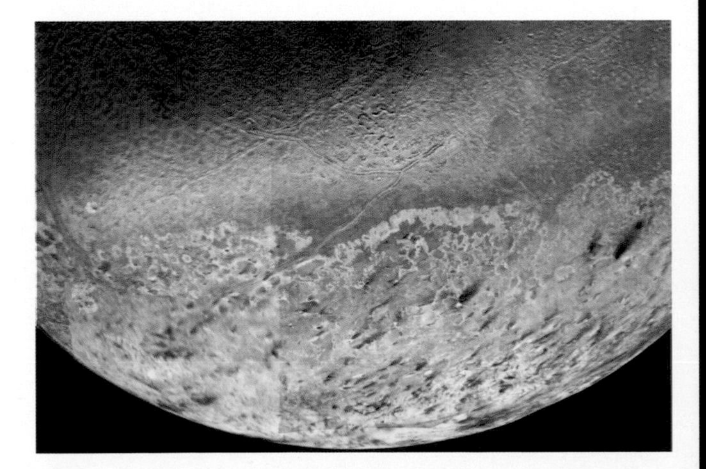

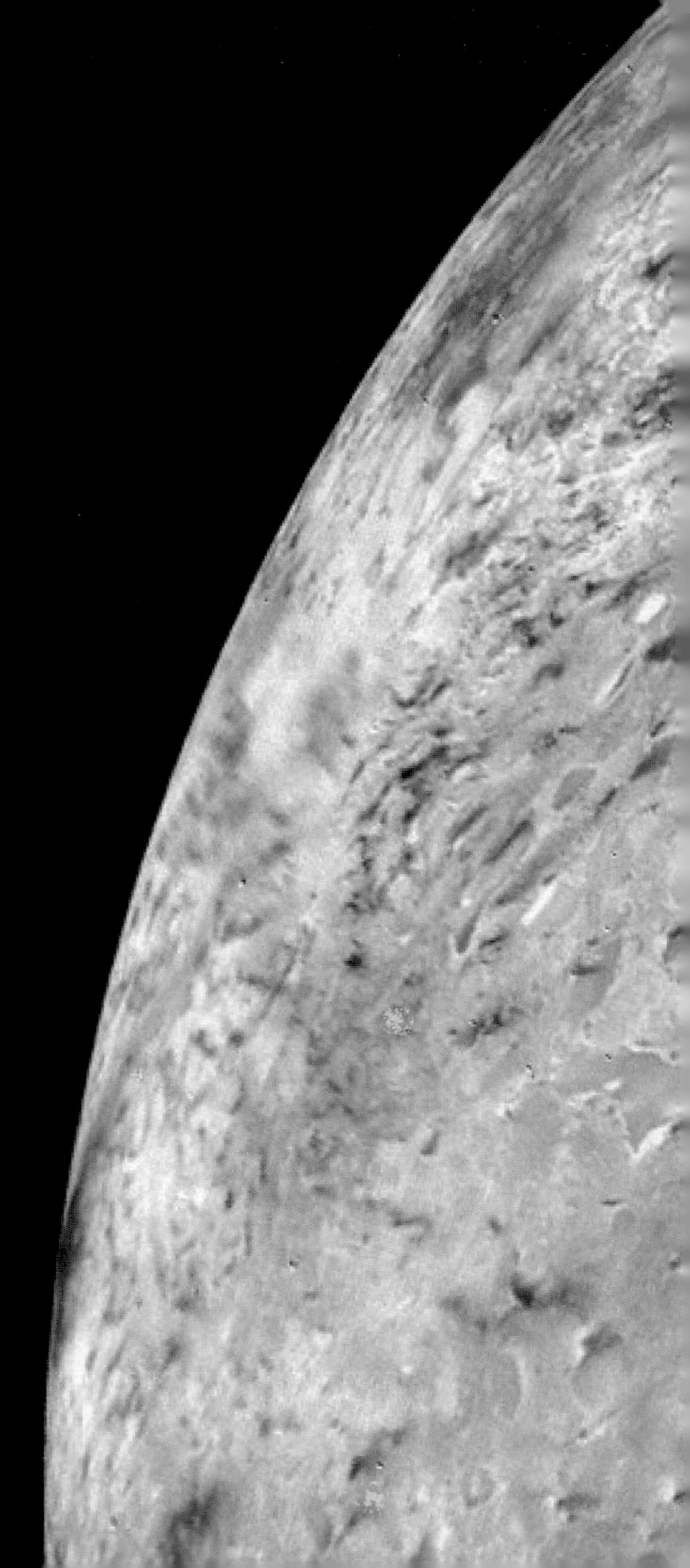

# Dwarf Planets, Asteroids & Comets

The Sun represents 99.85 per cent of the mass of the Solar System, whilst the eight planets make up a further 0.135 per cent. That leaves 0.015 per cent of the total material distributed between countless smaller bodies, mostly made up of rock and ice. Most of these occupy two regions in space: the asteroid belt between Mars and Jupiter (between two and four times Earth's distance from the Sun) and the Kuiper belt beyond Neptune (30 to 50 times Earth's distance from the Sun).

In the eighteenth century it was recognized that there is a large gap between Mars and Jupiter, wide enough to contain a planet. In 1801 astronomers declared they had found one, named Ceres. However, Ceres is far smaller than the planets and within a few decades several smaller objects were found in the same region of space. They were all classed as asteroids, which are sometimes referred to as 'minor planets'. Ceres has now been reclassified as a dwarf planet. The term 'dwarf planet' means it is large enough to be spherical but not big enough to have cleared its orbit of other objects.

In 1930, Pluto was discovered. Like Ceres more than a century earlier, it was initially proclaimed to be a planet. However, in the twenty-first century several more objects of similar size to Pluto have been discovered beyond Neptune. By 2006 it became apparent that there is a host of such tiny worlds in the icy reaches of the outer Solar System, which is now referred to as the Kuiper Belt. Accordingly, Pluto was downgraded to the status of dwarf planet.

Comets are small icy bodies that originate in the Kuiper Belt, the scattered disc around it and the much more distant Oort Cloud (situated more than 2000 times Earth's distance from the Sun). When comets are disturbed by the gravity of giant planets or nearby stars, they may enter the inner Solar System on highly elliptical orbits.

Dwarf planets, asteroids and comets are fascinating objects. Our understanding of them has advanced tremendously in the last two decades as several, including Ceres and Pluto, have been visited by spacecraft.

OPPOSITE:

**Comet NEOWISE**

In July 2020, Comet NEOWISE passed 103 million km (64 million miles) from Earth and was a spectacular sight in northern hemisphere night skies. Its orbit takes thousands of years to complete, so that the last time it was visible in Earth's night sky was around the time that Stonehenge was being built 4,400 years ago. As comets approach the Sun they are heated, which leads to the release of some of their material as gases. This is what makes them visible from Earth, especially if the comet forms a dust tail.

ALL:

**Halley's comet**

The best-known comet is named after astronomer Edmund Halley, who studied reports of a comet that approached Earth in 1531, 1607 and 1682, leading him to conclude they were the same comet reappearing at roughly 75 year intervals. He predicted its return in 1758, which proved correct. Halley's comet is on an elliptical orbit that carries it 35 times further than Earth from the Sun, beyond the orbit of Neptune. Its most recent visit to the inner Solar System was in 1986.

Several spacecraft were despatched to rendezvous with Halley's comet in 1986. The closest approach was by a probe named *Giotto*, after an Italian artist who saw the comet in 1301 and included it in his painting *Adoration of the Magi*. *Giotto* came within 600km (375 miles) of the comet's nucleus and found Halley's comet to be very dark, irregular in shape and 15km (9.3 miles) in length. The probe's multicolour camera was destroyed when it was hit by a dust particle but not before it returned an image of the nucleus (right). The comet is made mostly of water ice with a surface covering of dust; it has been described as a giant dirty snowball. Three outgassing jets were observed on the sunlit side of the comet, ejecting material that forms the comet's coma and tail.

LEFT & BELOW:

**Comet Tempel 1**

Tempel 1 is a short-period comet which returns every five and a half years. Its orbit is frequently perturbed by Jupiter's gravity, leading to changes in its shape and duration.

In 2005 the comet was deliberately struck by part of the probe *Deep Impact*, to observe the effect and discover what it is made of. Compounds detected in the dust raised by the impact included silicates, carbonates, smectite, sulphides and hydrocarbons. A subsequent visit by another spacecraft, *Stardust*, photographed the resulting crater and found it to be 150m (500ft) across.

Pictured left is an artist's impression of *Deep Impact* at Comet Tempel 1.

The picture below shows the moment of impact photographed by the fly-by section of *Deep Impact*.

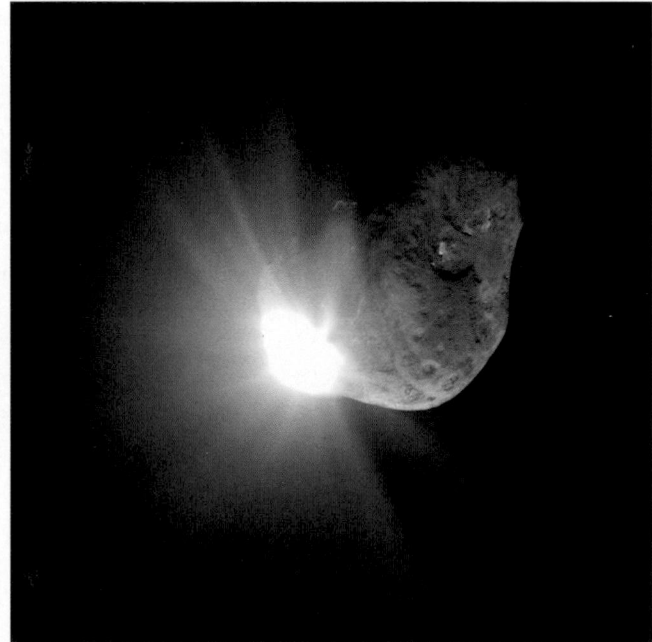

RIGHT:

**Comet Churyumov-Gerasimenko**

Circling the Sun in a relatively short orbit of six and a half years, Churyumov-Gerasimenko was the first comet on which a soft landing was successfully accomplished. A probe named *Rosetta* was launched in 2004, following which it made two fly-bys of Earth followed by one of Mars, using their gravity to accelerate and match its trajectory to that of the comet. It arrived in August 2014 and became the first spacecraft to orbit a comet. Its lander *Philae* successfully touched down on 12 November 2014. Unfortunately *Philae* bounced into the shadow of a cliff where its solar panels could not operate, so it ran out of power after two days.

*Rosetta* remained in orbit for two years and returned the best photographs ever made of a cometary nucleus. About 7km (4.3 miles) in length, Churyumov-Gerasimenko is a contact binary, meaning it comprises two objects that are touching but not structurally connected. *Rosetta* studied the changes that occurred as the comet approached the Sun, causing landslides and cliff collapses as its ice started to sublimate and outgassing occurred. Water vapour was analysed and found to contain a different isotope ratio to Earth's water, indicating that comets of this type were not the source of Earth's water. Numerous hydrocarbon compounds and molecular oxygen were detected, showing that these building blocks of life were present in the early Solar System.

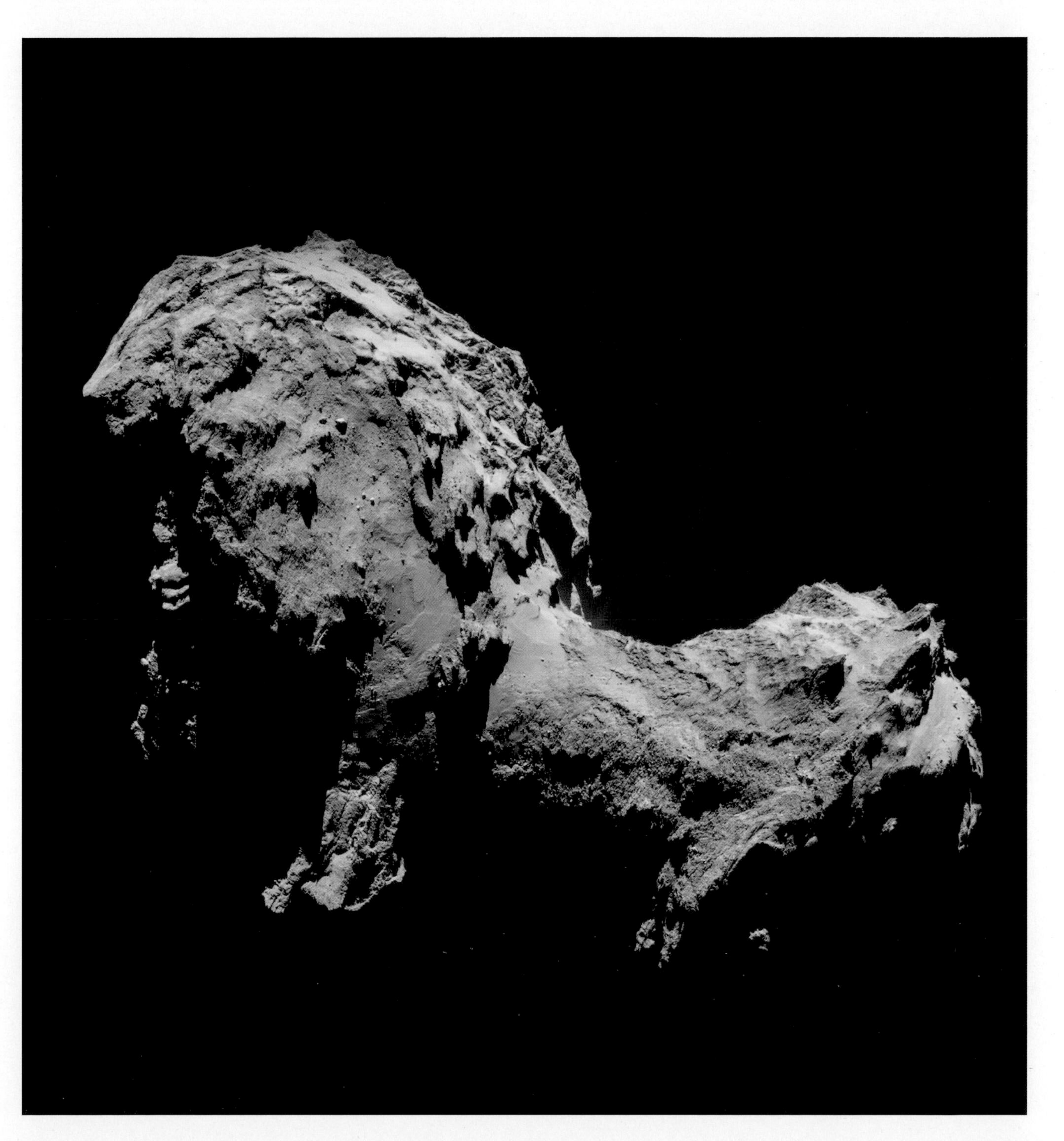

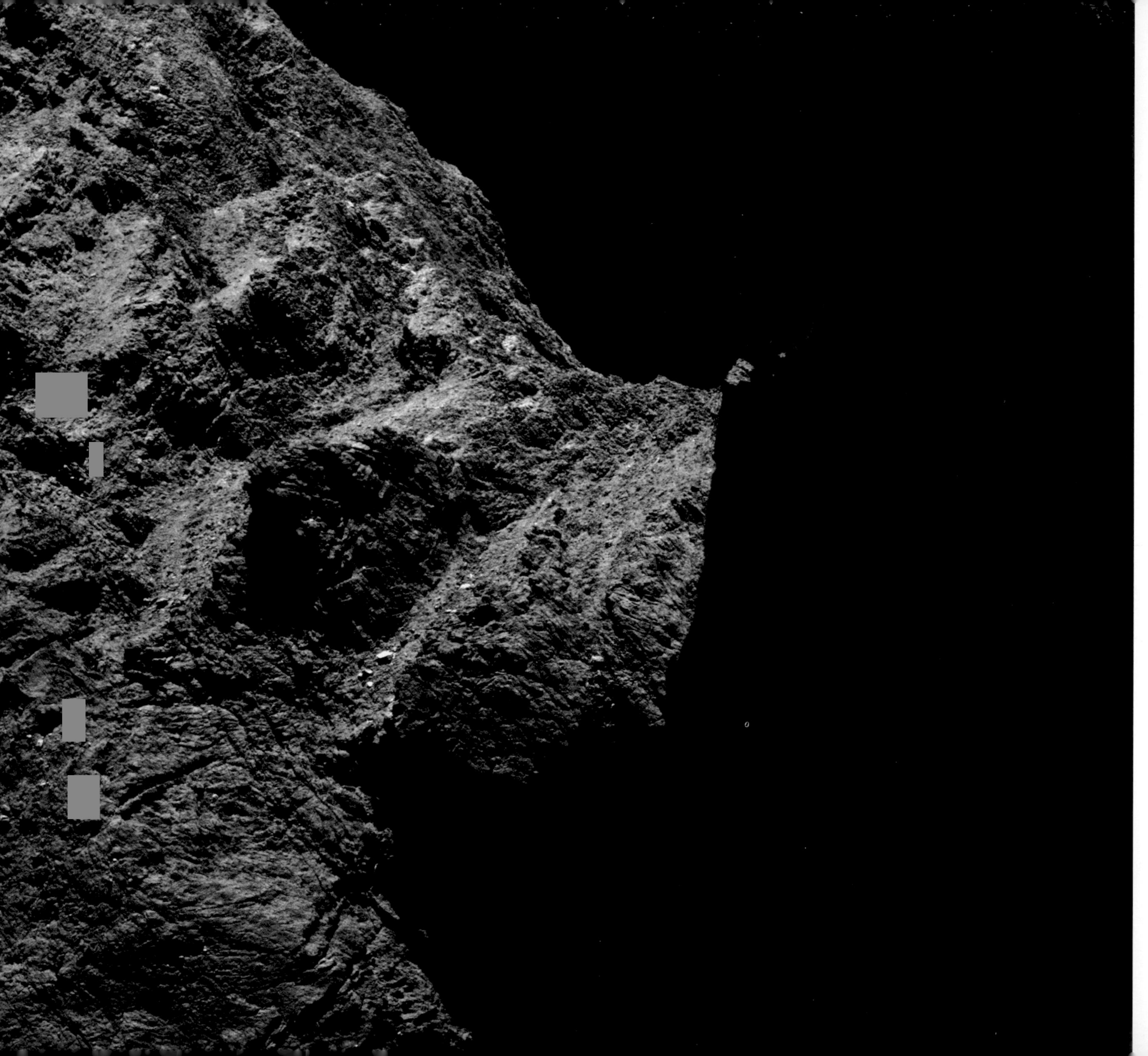

LEFT:

**Night and day**

A close-up view of the boundary between night and day on Comet Churyumov-Gerasimenko, taken from *Rosetta*.

OPPOSITE:

**Gaspra**

Located close to the inner edge of the asteroid belt beyond Mars, Gaspra was the first asteroid to be photographed by a spacecraft. *Galileo* passed within 1600km (1000 miles) on 29 October 1991 en route to Jupiter. The asteroid rotates every seven hours, so *Galileo* was able to image 80 per cent of the surface as it flew by.

Gaspra is 18km (11 miles) along its longest axis, making it intermediate in size between Mars's moons Phobos and Deimos. The surface is smoother with fewer large craters than expected and a number of flat areas, giving it a notably angular appearance. It is thought that Gaspra formed within the last 300 million years as a result of a collision breaking up a larger asteroid.

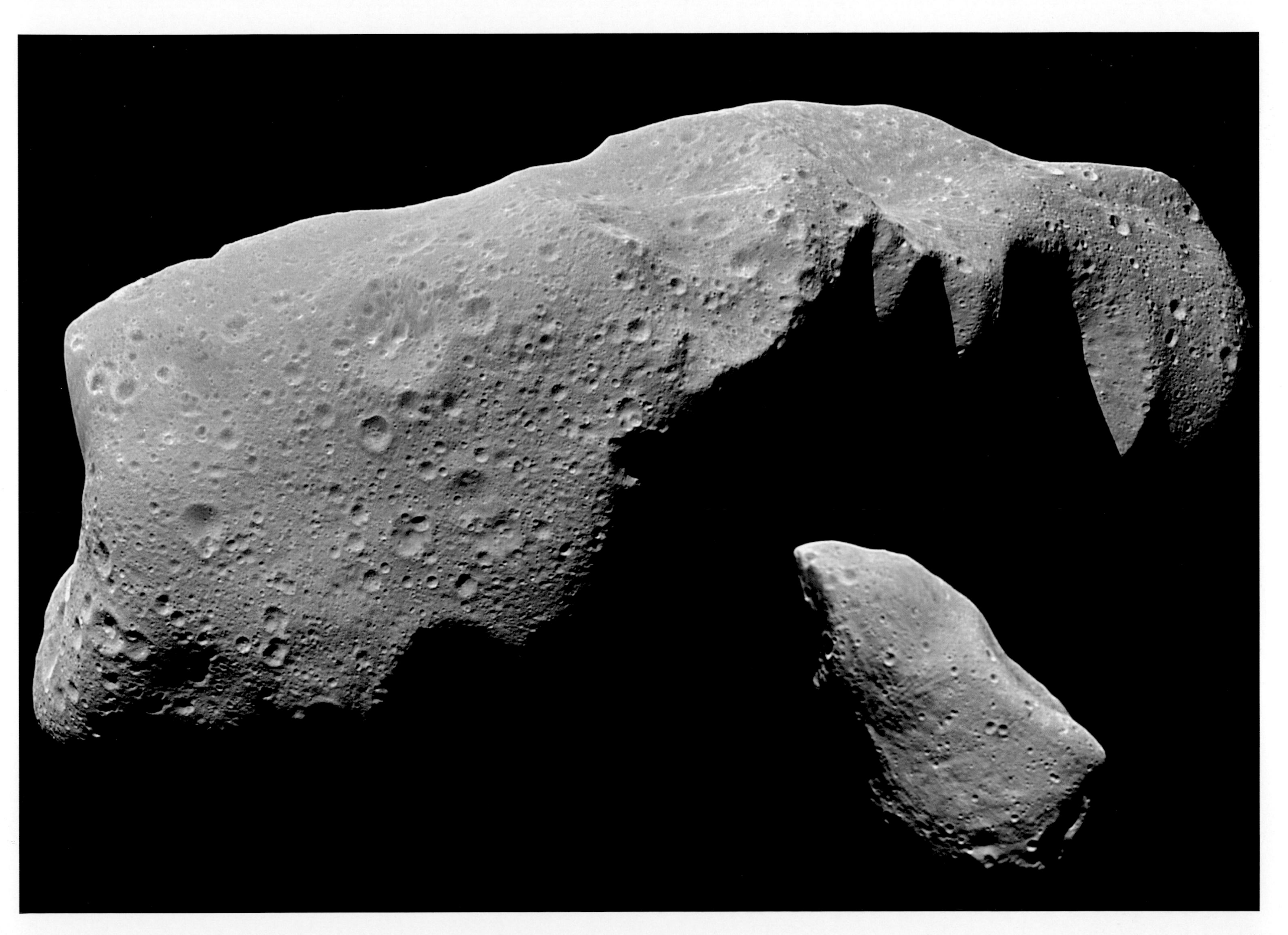

ABOVE:
**Eros**
Eros, the first asteroid landed on by a spacecraft, is of particular interest because its orbit crosses that of Mars and brings it relatively close to Earth. There is a theoretical possibility that millions of years in the future its orbit could evolve to cross that of Earth. The probe *NEAR* (*Near Earth Asteroid Rendezvous*) *Shoemaker* flew by Eros in 1998, orbited it in 2000 and landed on the surface on 12 February 2001.

Eros is 34 km (21 miles) and somewhat like the shape of a peanut. Its surface was pounded by a large impact around one billion years ago that left a large crater and created seismic shockwaves across the asteroid, filling up many smaller craters with rubble. *NEAR Shoemaker* found evidence that Eros may be rich in aluminium and possibly rare metals such as gold and platinum.

ABOVE:
**Bennu**
Another near-Earth asteroid, Bennu, has an orbit which crosses that of our planet. Its current path around the Sun takes 1.2 years but is frequently perturbed as a result of gravitational interactions. The closest Earth currently comes to Bennu's orbit is 480,000km (300,000 miles) – a little more than the distance to the Moon – between 23 and 25 September each year.

Bennu is categorized as a potentially hazardous object, with a one in 1800 risk of striking Earth between 2178 and 2290. Measuring 565m (1850ft) across its longest axis, Bennu is big enough to cause devastation over hundreds of kilometres in the event of an impact. Were such a collision to be predicted, it is likely that efforts would be made to deflect the asteroid.

Following its discovery in 1999, Bennu was visited by *OSIRIS-REx*, which arrived in 2018. On 20 October 2020, the spacecraft collected samples for return to Earth in September 2023. Analyses of these will improve our understanding of the asteroid's composition. Because Bennu is effectively a leftover building block of the planets, it will tell us more about the formation of the Solar System and the source of organic compounds on Earth.

ABOVE:
**Lutetia**
Located in the main asteroid belt, Lutetia is 120km (74 miles) across its longest axis. It was photographed by *Rosetta* from a distance of 3162km (1964 miles), showing surface features down to 60m (190ft) in size in its northern hemisphere. The asteroid is unusually dense and may consist of metal-enriched silicates.

ABOVE:
**Vesta**
With a length of 572km (355 miles), Vesta is the second-largest asteroid and the brightest as seen from Earth. It was the first destination of *Dawn*, which arrived in September 2011 and spent a year in orbit. Located in the main asteroid belt between Mars and Jupiter, Vesta takes 3.6 years to orbit the Sun and 5.3 hours to rotate on its axis. It is almost large enough to be spherical as a result of its own gravity and is thought to be differentiated into a crust, mantle and metallic core. The most prominent surface features are two huge impact craters near its south pole. With diameters of 400km (250 miles) and 500km (310 miles), they approach the diameter of the asteroid in size. Rock fragments blasted off the surface by these impacts is thought to fall to Earth as HED meteorites. Three prominent craters in the northern hemisphere, visible in this image, are nicknamed 'snowman' because of the way they overlap.

Despite its small size, Vesta has some of the largest chasms in the Solar System, longer and wider than the Grand Canyon. These troughs sculpt much of the equatorial region and are visible at the top of this image. They are believed to be graben, formed by faulting induced by collision with another asteroid.

ABOVE:
**Ceres**
Ceres is the largest member of the asteroid belt, 964km (590 miles) in diameter. It is the only such object large enough to be spherical and it therefore qualifies as a dwarf planet. One orbit of the Sun takes 4.6 years and a Cererian day lasts 9 hours 4 minutes. It was the second destination visited by *Dawn*, which orbited the dwarf planet between March 2015 and November 2018. In a series of orbits at different altitudes, Dawn photographed and mapped the entire surface.

One of the most surprising findings by *Dawn* was evidence of recent cryovolcanism or outgassing on the surface of Ceres. Bright spots within a crater appear to be brine rich in sodium carbonate extruded from the interior. A relative paucity of large craters suggest resurfacing by internal geological activity, which has waned over time as the dwarf planet cooled.

Another discovery was the presence of ammonia salts within Occator Crater. It is thought that Ceres most likely formed beyond the orbit of Jupiter, where ammonia is more abundant, and migrated inwards to the asteroid belt as a result of gravitational interactions with the gas giant. Pictured above is the northern hemisphere of Ceres, viewed from 22,000km (14,000 miles) above the north pole. A bright spot indicating extrusion of brine is visible in the crater on the right.

**Pluto**

On 19 January 2006, *New Horizons* was launched to what was then considered to be the most distant planet and (since 1989) the only one not yet visited by a spacecraft from Earth. Later that year, the International Astronomical Union adopted a resolution to define a planet and Pluto was demoted to the status of dwarf planet. So when *New Horizons* arrived nine and a half years after launch, its destination was no longer considered a planet. The reason is that although Pluto is spherical and orbits the Sun, it is not big enough to gravitationally dominate its orbit.

In fact, Pluto's 248-year orbit is highly eccentric, overlapping with that of Neptune. Between 1979 and 1999, the dwarf planet was closer to the Sun than the ice giant. The two bodies are in a stable orbital resonance with each other, Neptune completing exactly three orbits for every two of Pluto, so they will never collide.

***New Horizons***

*New Horizons* was the fastest spacecraft ever launched. Travelling at over 16km/second (10 miles/sec) from Earth, it was already going fast enough to escape the Sun's gravity and leave the Solar System. A year later it passed 2.3 million km (1.4 million miles) from Jupiter, from which it received a gravity assist that accelerated its speed to 23km/second (14.2 miles/sec). That boost cut three years from the time it took *New Horizons* to reach the dwarf planet Pluto.

Eight and a half years later, on 14 July 2015, *New Horizons* passed 13,000km (8000 miles) from Pluto as shown in this artist's impression. It took four hours and 25 minutes for radio signals to travel back to Earth relaying the success of this extraordinary encounter, 4.7 billion km (2.9 billion miles) from home. Transmitting across this incredible distance, *New Horizons* spent the next 16 months telling us everything it had found.

**Studying Pluto**

Pluto is much smaller than the planets. Its diameter of 2377km (1476 miles) is half that of Mercury and less than seven of the Solar System's moons. Pluto's mass is just one-sixth that of Earth's Moon and a little less than the mass of Eris, another dwarf planet beyond Neptune that was discovered in 2005. It is thought there may be hundreds of objects of similar size to Pluto in the Kuiper Belt and scattered disc beyond Neptune.

Internally, Pluto comprises rock and water ice. These are likely to be differentiated, with rock in the core and a mantle of ice. Internal radioactivity may provide sufficient heat to melt some of the ice into a subsurface ocean. There is no magnetic field.

Pluto's surface is surprisingly varied and colourful, with mountains made of water ice and plains of nitrogen ice. One of the most distinctive features is a large, bright, heart-shaped area called Tombaugh Regio (after Clyde Tombaugh, who discovered Pluto in 1930). Its western region, Sputnik Planitia, is a 1000km(620 mile)-wide basin of frozen nitrogen and carbon monoxide. The basin is divided into polygonal structures, thought to be blocks of water ice behaving like tectonic plates on Earth. There is also evidence of recent glacial activity. The absence of craters indicates the visible surface of Sputnik Planitia is only a few hundred thousand years old, implying resurfacing by active geological processes.

Elsewhere, *New Horizons* photographed what look like volcanoes. With surface temperatures as low as -240°C (-400°F), water ice would behave like magma if it melted and erupted through the surface. Such features are termed cryovolcanoes.

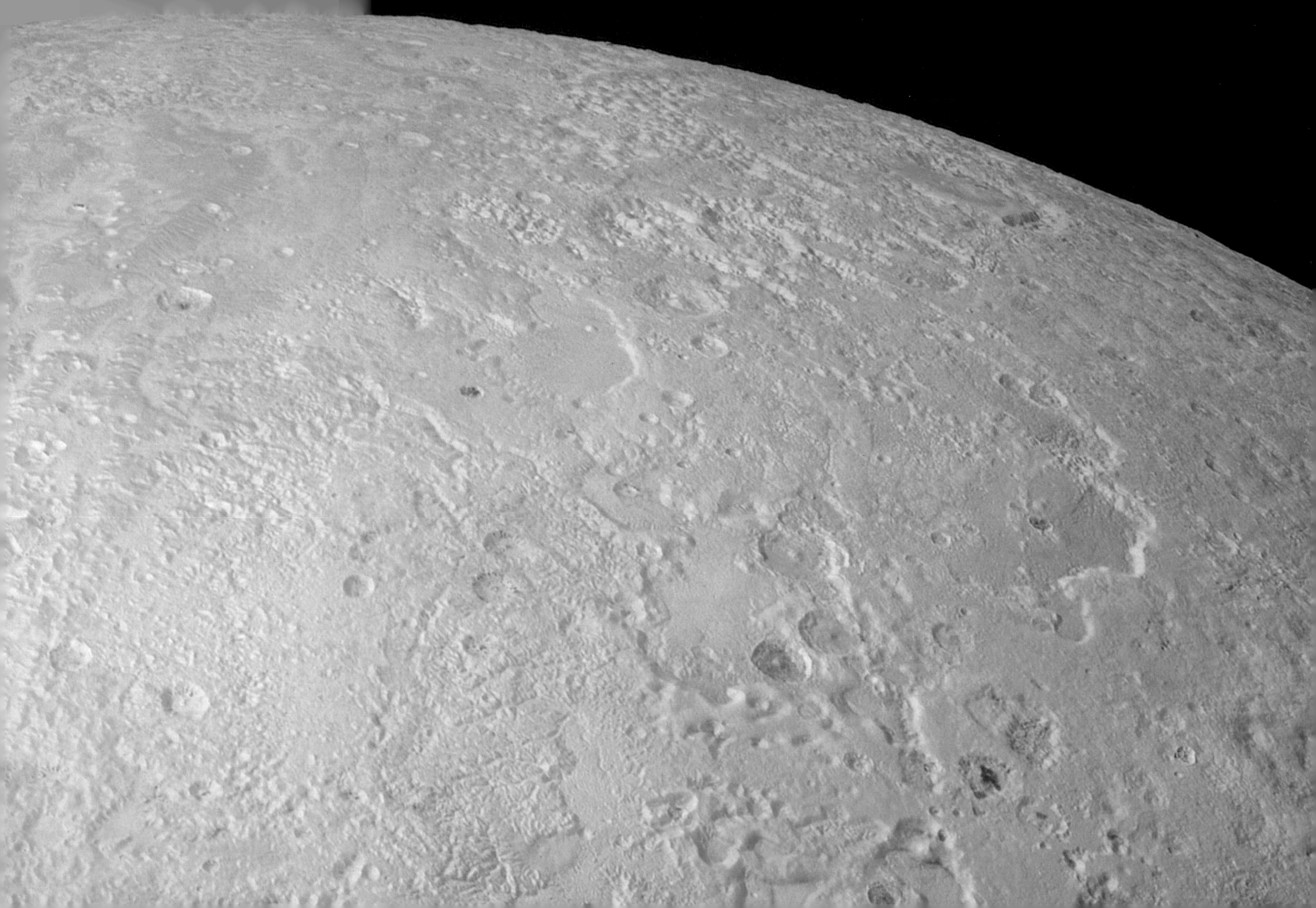

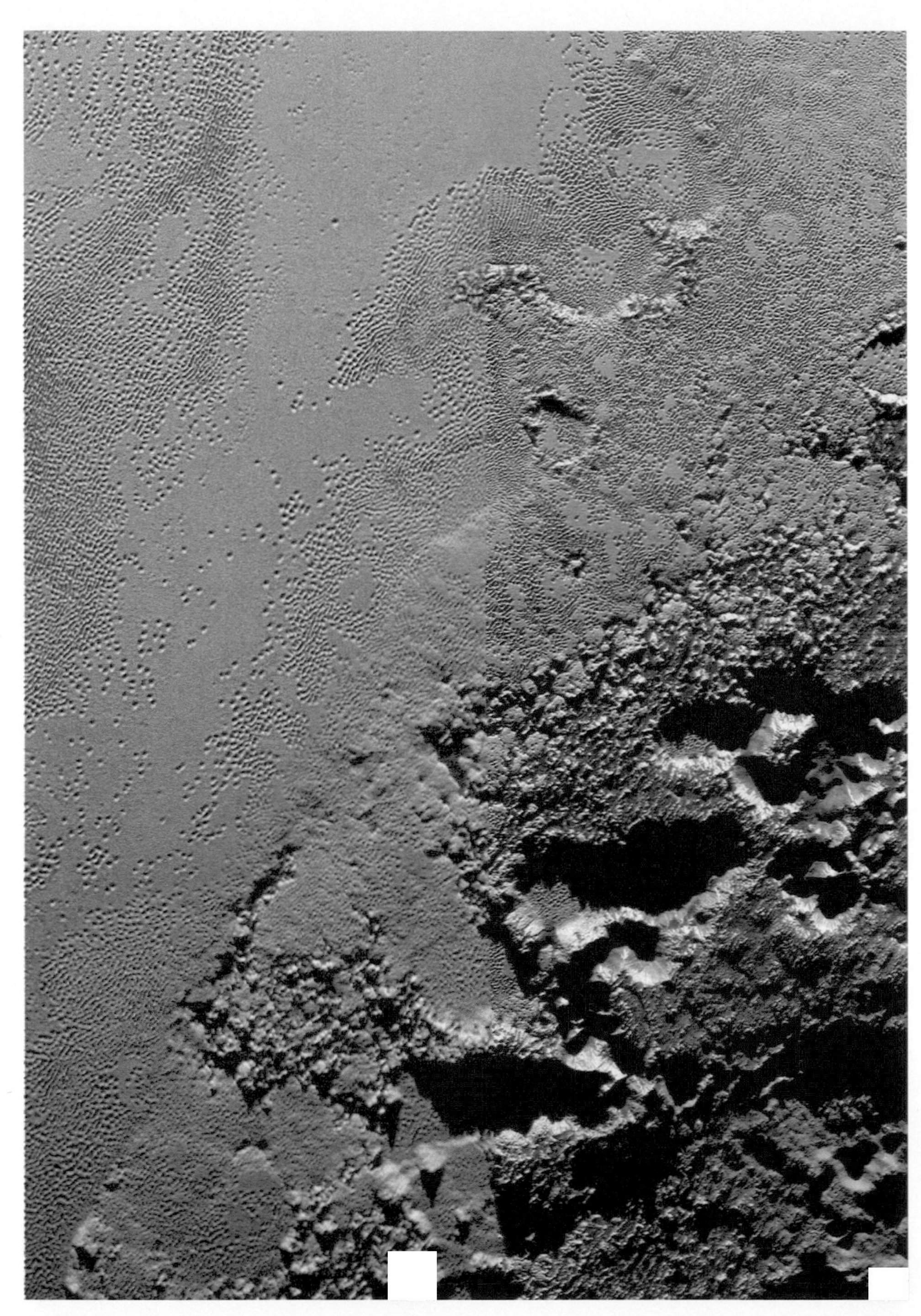

LEFT:

**Surface of Pluto**

This image shows the boundary between plains of nitrogen ice and dark highlands of water ice, thought to derive their red colour from tholin molecules.

BELOW:

**Sputnik Planitia**

Pluto has a tenuous atmosphere of nitrogen, methane and carbon monoxide, with a pressure 100,000 times lower than on Earth. Nonetheless, in Sputnik Planitia, dunes have been formed by winds blowing towards nearby mountains. The 'sand' in these dunes is likely to be particles of solid methane. As Pluto's elliptical orbit carries it even further from the Sun, atmospheric pressure is falling because its gases are freezing onto the surface.

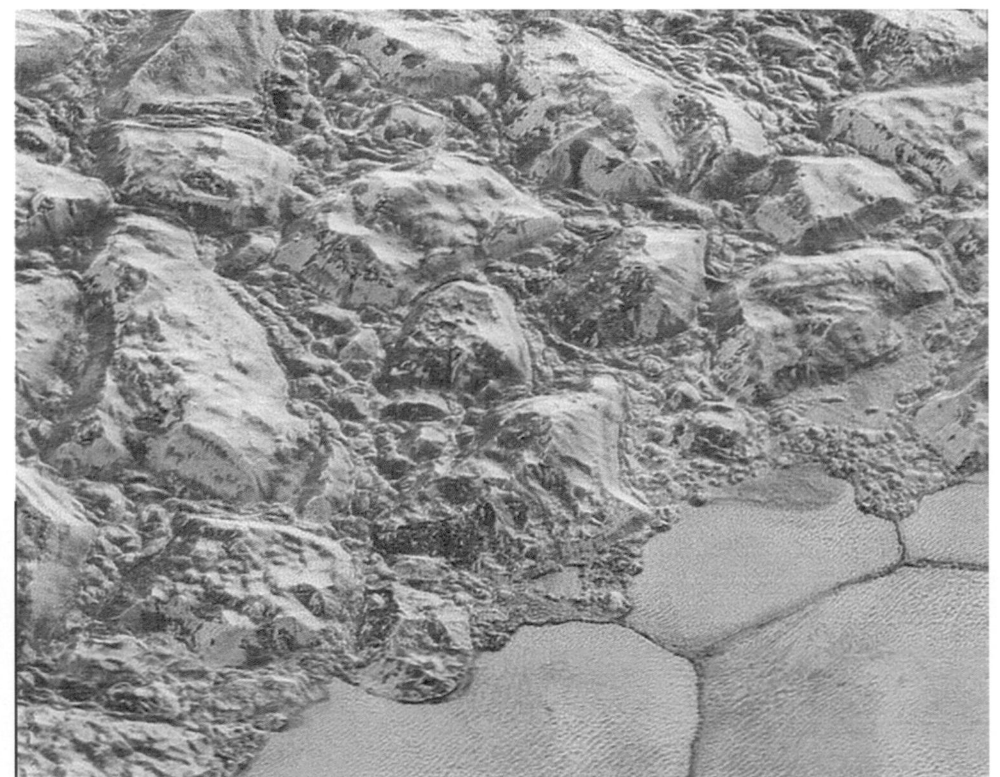

**Pluto and Charon**

Curiously for such a small world, Pluto has no fewer than five moons. All the moons, in common with Pluto itself, are named after mythological beings associated with the underworld. Four of them (Styx, Nix, Kerberos and Hydra) are less than 50km (30 miles) wide and irregular in shape. It is thought that, like Earth's Moon, they formed from fragments thrown out in a giant collision of Pluto with another object.

The largest moon, Charon, is over half the diameter of Pluto. In fact, Pluto and Charon form a binary system orbiting a common centre of mass between them. They are sometimes described as a double dwarf planet. Both Pluto and Charon are tidally locked to each other, with orbital periods and rotations of 6.4 days. Pluto can only be seen from one hemisphere of Charon; whilst Charon can only be seen from one hemisphere of Pluto.

Pictured is a mosaic of Pluto and Charon shown to the same scale (the distance between them is not to scale).

ALL:

**Charon**

Charon is 20,000km (12,500 miles) from Pluto, a distance covered by *New Horizons* in just 15 minutes. With a diameter of 1212km (750 miles) it is the largest known moon relative to its parent body. Like Pluto, Charon comprises rock and water ice, probably separated into a rocky core and an icy mantle. Charon's surface is somewhat different to Pluto, appearing to be dominated by water ice. The presence of water crystals and ammonia hydrates on the surface suggests that cryovolcanoes erupting molten ice may be active. Charon's poles are darker than its equatorial regions. They are believed to result from seasonal deposition of frozen nitrogen, carbon monoxide and methane at winter temperatures of -258°C (-432°F). These compounds subsequently react in spring sunlight to form complex organic molecules called tholins, which are reddish in colour. Other surface features include troughs extending 1000km (600 miles) in length, which are likely to be graben resulting from faulting of the icy crust.

Pictured left is Charon, showing the red Mordor Macula region at its north pole. Pictured right is a topographically enhanced view of Charon using data returned by *New Horizons*.

RIGHT:

### Goodbye to Pluto

As *New Horizons* departed Pluto, it took this view of the dwarf planet backlit by the distant Sun. Part of Sputnik Planitia can be seen in the sunlit crescent at the top. Haze around the circumference of the disc is Pluto's atmosphere.

OVERLEAF:

### Pits on Pluto

*New Horizons* took the highest resolution images ever obtained of the intricate pattern of 'pits' across a section of Pluto's prominent heart-shaped region, informally named Tombaugh Regio. Mission scientists believe these mysterious indentations may form through a combination of ice fracturing and evaporation. The scarcity of overlying impact craters in this area also leads scientists to conclude that these pits – typically hundreds of metres across – formed relatively recently. Their alignment provides clues about the ice flow and the exchange of nitrogen and other volatile materials between the surface and the atmosphere.

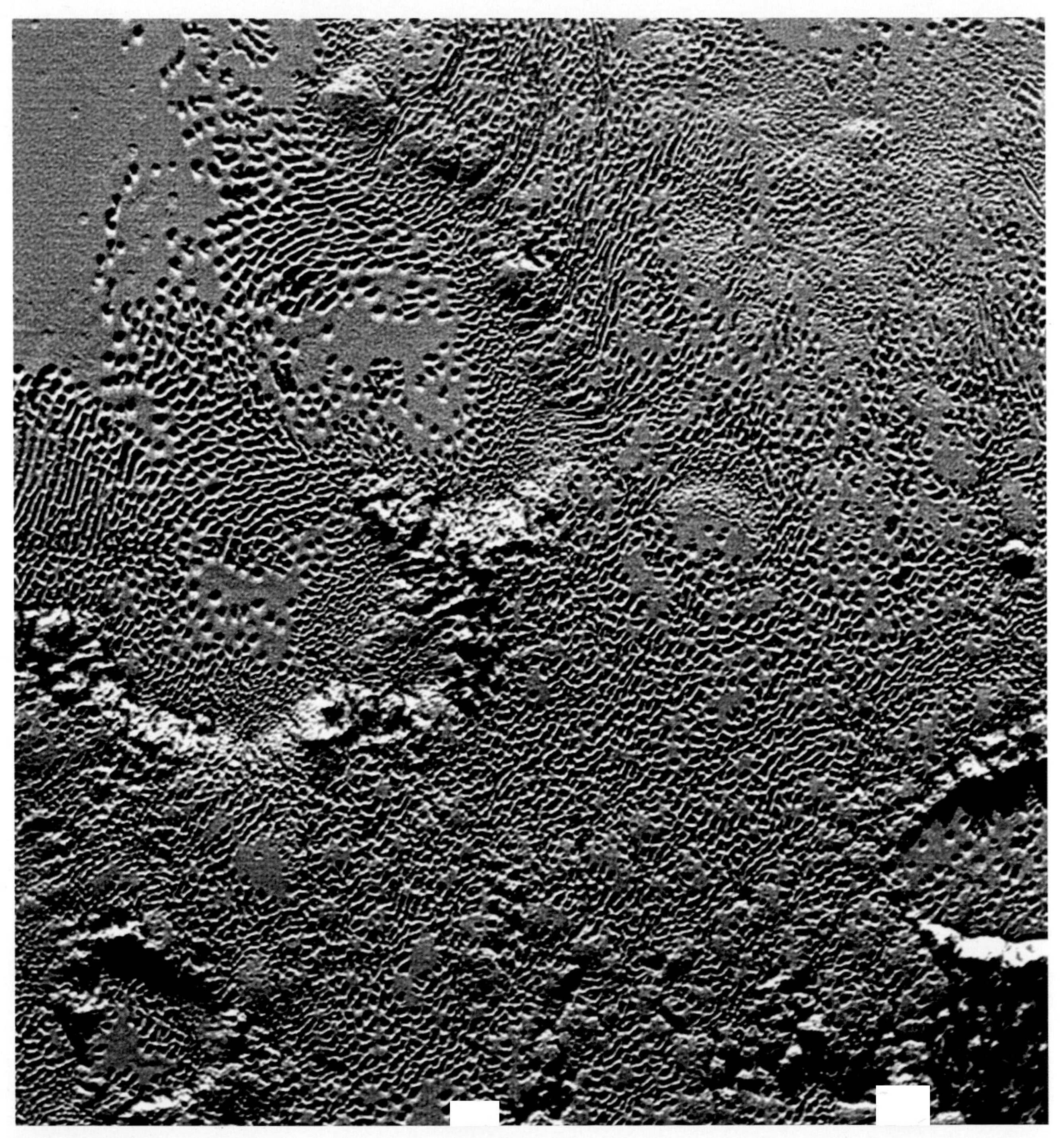

RIGHT:

**Arrokoth**

After leaving Pluto, *New Horizons* was not retired. It went on to visit another Kuiper Belt object called Arrokoth on 1 January 2019. Arrokoth is a contact binary comprising two icy planetismals stuck together to look somewhat like a snowman. It is 36km (22 miles) in length and distinctly reddish in colour due to surface tholins. Largely unaltered since the early days of the Solar System, it gives a hint of what the original building blocks of the planets may have been like. The superficial resemblance of Arrokoth to Comet Churyumov-Gerasimenko may not be coincidental, for the Kuiper belt is where some comets originate.

When *New Horizons* passed by at a distance of 3500km (2200 miles), Arrokoth was given the nickname 'Ultima Thule', meaning a far distant land. It is 6.7 billion km (4.2 billion miles) from the Sun, which is 45 times further than Earth, and takes 298 years to complete a single orbit. This is the most distant object yet visited in our exploration of the Solar System.

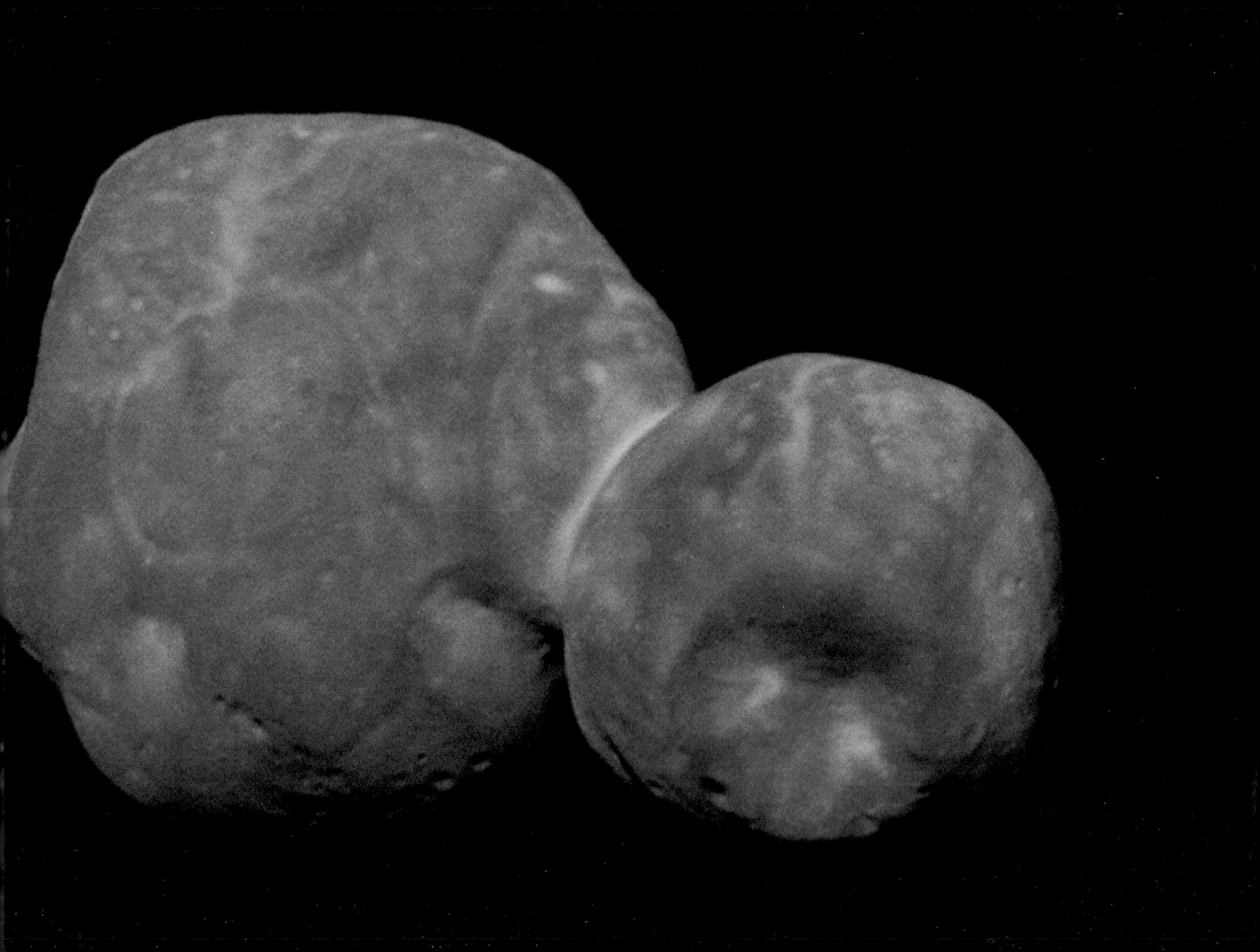

# Appendix: Solar System exploration missions mentioned in the text

| Name | Space agency | Year launched | Mission ended | Destination |
|---|---|---|---|---|
| Akatsuki | JAXA | 2010 | ongoing | Venus (orbiter) |
| Apollo 8 (manned) | NASA | 1968 | 1968 | Moon (orbiter) |
| Apollo 11 (manned) | NASA | 1969 | 1968 | Moon (landing) |
| Apollo 12 (manned) | NASA | 1969 | 1969 | Moon (landing) |
| Apollo 17 (manned) | NASA | 1972 | 1972 | Moon (landing) |
| BepiColombo | ESA and JAXA | 2018 | ongoing | Mercury (orbiter) |
| Cassini-Huygens | NASA, ESA, ASI | 1997 | 2017 | Saturn (orbiter), Titan (lander) |
| Curiosity | NASA | 2011 | ongoing | Mars (rover) |
| Dawn | NASA | 2007 | 2018 | Vesta (orbiter), Ceres (orbiter) |
| Deep Impact | NASA | 2005 | 2013 | Comet Tempel 1, Comet Hartley 2 |
| Deep Space Climate Observatory | NASA and NOAA | 2015 | ongoing | Earth |
| Dragonfly | NASA | 2027* | - | Titan (lander) |
| Europa Clipper | NASA | 2024* | - | Jupiter (orbiter), Europa |
| Galileo | NASA | 1989 | 2003 | Jupiter (orbiter) |
| Giotto | ESA | 1985 | 1992 | Halley's Comet, Comet Grigg-Skjellerup |
| Hinode | JAXA | 2006 | ongoing | Earth orbit (solar observing) |
| Hubble Space Telescope | NASA | 1990 | ongoing | Earth orbit (distant observing) |
| International Space Station (manned) | NASA, Roscosmos, JAXA, ESA and CSA | 1998 | ongoing | Earth orbit |
| Juno | NASA | 2011 | ongoing | Jupiter (orbiter) |
| Jupiter Icy Moons Explorer (JUICE) | ESA | 2023* | - | Ganymede (orbiter) |
| Luna 3 | CCCP | 1959 | 1959 | Moon |
| Mariner 2 | NASA | 1962 | 1963 | Venus |
| Mariner 4 | NASA | 1964 | 1967 | Mars |
| Mariner 9 | NASA | 1971 | 1972 | Mars (orbiter) |
| Mariner 10 | NASA | 1973 | 1975 | Venus, Mercury |
| Magellan | NASA | 1989 | 1994 | Venus (orbiter) |
| Mars Express | ESA | 2003 | ongoing | Mars (orbiter) |
| Mars Global Surveyor | NASA | 1996 | 2007 | Mars (orbiter) |
| Mars Odyssey | NASA | 2001 | ongoing | Mars (orbiter) |

| Name | Space agency | Year launched | Mission ended | Destination |
|---|---|---|---|---|
| Mars Reconnaissance Orbiter | NASA | 2005 | ongoing | Mars (orbiter) |
| Messenger | NASA | 2004 | 2015 | Mercury (orbiter) |
| NEAR Shoemaker | NASA | 1996 | 2001 | Eros (lander) |
| New Horizons | NASA | 2006 | ongoing | Jupiter, Pluto, Arrokoth |
| NuSTAR (Nuclear Spectroscopic Telescope Array) | NASA | 2012 | ongoing | Earth orbit (solar observing) |
| Opportunity | NASA | 2003 | 2018 | Mars (rover) |
| OSIRIS-REx | NASA | 2016 | ongoing | Asteroid Bennu (lander) |
| Parker Solar Probe | NASA | 2018 | ongoing | The Sun (orbiter) |
| Perseverance and Ingenuity | NASA | 2020 | ongoing | Mars (rover and helicopter) |
| Phoenix | NASA | 2007 | 2008 | Mars (lander) |
| Pioneer 10 | NASA | 1972 | 2003 | Jupiter |
| Pioneer 11 | NASA | 1973 | 2003 | Jupiter, Saturn |
| Rosetta and Philae | ESA | 2004 | 2016 | Asteroids Steins and Lutetia, Comet Churyumov-Gerasimenko (orbiter and lander) |
| Solar Dynamics Observatory | NASA | 2010 | ongoing | Earth orbit (solar observing) |
| Spirit | NASA | 2003 | 2010 | Mars (rover) |
| Stardust | NASA | 1999 | 2011 | Comet Wild 2, Asteroid Annefrank, Comet Tempel 1 |
| Tainwen-1 and Zhurong | CNSA | 2020 | ongoing | Mars (lander & rover) |
| Venera 7 | CCCP | 1970 | 1970 | Venus (lander) |
| Venera 10 | CCCP | 1975 | 1975 | Venus (lander) |
| Venera 13 | CCCP | 1981 | 1982 | Venus (lander) |
| Venera 14 | CCCP | 1981 | 1982 | Venus (lander) |
| Venus Express | ESA | 2005 | 2014 | Venus (orbiter) |
| Venus Pioneer | NASA | 1978 | 1992 | Venus (orbiter & lander) |
| Viking 1 | NASA | 1975 | 1982 | Mars (orbiter & lander) |
| Viking 2 | NASA | 1975 | 1980 | Mars (orbiter & lander) |
| Voyager 1 | NASA | 1978 | ongoing | Jupiter, Saturn |
| Voyager 2 | NASA | 1978 | ongoing | Jupiter, Saturn, Uranus, Neptune |

* indicates missions not yet launched at time of going to press

**Key to space agencies:**

| | |
|---|---|
| ASI | Italian Space Agency |
| CCCP | Soviet Space Programme |
| CNSA | China National Space Administration |
| CSA | Canadian Space Agency |
| ESA | European Space Agency |
| JAXA | Japan Aerospace Exploration Agency |
| NASA | United States National Aeronautics and Space Administration |
| NOAA | United States National Oceanic and Atmospheric Administration |
| Roscosmos | Russian Space Agency |

# Picture Credits

ABBREVIATIONS: '*NASA/JPL-Caltech*' is credited as 'N/J-C'; '*NASA/Johns Hopkins University Applied Physics Laboratory*' is credited as 'N/JHUAPL'

2: NASA; 5: N/J-C; 6: Steven Hobbs/Stocktrek Images/Getty Images; 7: N/J-C/Space Science Institute; 8: N/J-C/ASU/MSSS; 9: NASA/Bill Ingalls; 10-11: NASA; 12/13: Dave Jarvis (licensed under the Creative Commons Attribution-Share Alike 3.0 Unported Licence); 14: ESA/NASA/SOHO; 16: NASA/SDO/AIA; 17: ESA/CESAR; 18: N/J-C; 19: NASA/Goddard; 20: ESA/ESAC/CESAR – A de Burgos; 21: N/J-C/GSFC/JAXA; 22 & 23: NASA/Goddard; 24/25: N/JHUAPL/Carnegie Institution of Washington; 26 & 27: NASA/Goddard; 28/29: a. v. ley/Getty Images; 30: NASA/JHUAPL; 32: NASA/Bill Ingalls; 33: N/J-C/USGS; 34-41: N/JHUAPL/Carnegie Institution of Washington; 42: NASA; 43: N/J-C; 44: NASA; 45-49: N/J-C; 50: JAXA/ISAS/DARTS/Kevin M Gill (licensed under the Creative Commons Attribution 2.0 Generic Licence); 51: ESA/C Carreau; 52: NASA; 54: BlueOrange Studio/Shutterstock; 55: Wim Hoek/Shutterstock; 56: Alex Mustard/Nature Picture Library/Alamy; 57: ESA (CC BY-SA 3.0 IGO); 58/59: Kastianz/Shutterstock; 60: Marc Stephan/Shutterstock; 61: Pablo Hidalgo/Dreamstime; 62: SRStudio/Shutterstock; 63: NASA; 64/65: Robert Harvey; 66/67: Dominic Jeanmaire/Shutterstock; 68/69 & 70: NASA; 71: NASA Earth Observatory/Joshua Stevens/NOAA National Environmental Satellite, Data & Information Service; 72/73: Gustavo Frazao/Shutterstock; 74/75: Willem Tims/Shutterstock; 76: N/J-C; 77: N/J-C/KSC; 78 & 79 left: NASA/Goddard; 79 right: Stocktrek Images/Alamy; 80: ESO/R Lucchesi; 81: Kertu/Shutterstock; 82: Alan Dyer/VWPics/Alamy; 83: NASA/ESA; 84: N/J-C; 85: James Stuby based on NASA image; 86: N/J-C; 87: James Stuby based on NASA image; 88/89: Malcolm Park/Alamy; 90-97: N/J-C; 98: NASA/USGS; 100: N/J-C; 101: NASA/Hubble Heritage Team/STScI/AURA; 102: N/J-C; 103: N/J-C/USGS; 104 top: ESA/DLR/FU Berlin/NASA MGS MOLA Science Team; 104 bottom: ESA/DLR/FU Berlin/Bill Dunford; 105-107: N/J-C/Univ of Arizona; 108: N/J-C/USGS; 109: ESA/DLR/FU Berlin (CC BY-SA 3.0 IGO); 110: ESA/DLR/Freie Univ Berlin (G Neukum); 111: N/J-C/Univ of Arizona; 112/113: N/J-C/USGS; 114 & 115: ESA/DLR/FU Berlin/G Neukum (CC BY-SA 3.0 IGO); 116 & 117: N/J-C/Univ of Arizona; 118: N/J-C/MSSS; 119: N/J-C/Univ of Arizona; 120: N/J-C/Malin Space Science Systems; 121: N/J-C/Univ of Arizona; 122 & 122/123: N/J-C/MSSS; 123 top, 124 & 125: N/J-C/Univ of Arizona; 126: N/J-C/MSSS/JHU-APL; 127 & 128/129: ESA/DLR/FU Berlin/G Neukum (CC BY-SA 3.0 IGO); 130: NASA/USGS; 131: N/J-C; 132/133: N/J-C/MSSS; 134/135: N/J-C/Cornell Univ/Arizona State Univ; 136 & 137: N/J-C/MSSS; 136/137: N/J-C/Cornell; 138: N/J-C; 139 & 140: N/J-C/MSSS; 141: NASA/GSFC; 142 & 143: N/J-C/Univ of Arizona; 144: Enhanced image by Kevin M Gill (CC-BY) based on images provided courtesy of N/J-C/SwRI/MSSS; 146: N/J-C; 147: NASA/ESA/A Simon (GSFC); 148: Enhanced image by Gerald Eichstädt based on images provided courtesy of N/J-C/SwRI/MSSS; 149: NASA/ESA/J Nichols (Univ of Leicester); 150: N/J-C/Cornell Univ; 151: NASA; 152 left: N/J-C/DLR; 152 right & 153: N/J-C/Univ of Arizona; 154: N/J-C/DLR; 155: N/J-C/SETI Institute; 156: N/J-C/Brown Univ; 157: NASA/JPL/SwRI/MSSS/Kevin M Gill; 158: N/J-C/ASU; 159: Getty Images; 160: N/J-C/SSI/Cornell Univ; 162: NASA/ESA/J Clarke (Boston Univ) & Z Levay (STScI); 163: N/J-C; 164-171: N/J-C/Space Science Institute; 172/173: Mark Garlick/Science Photo Library/Getty Images; 174 left: N/J-C/Space Science Institute; 174 right: N/J-C/Univ of Arizona/Univ of Idaho; 175: N/J-C/Univ of Nantes/Univ of Arizona; 176 left: N/J-C/ASI/Cornell; 176 right: N/J-C/ESA/Univ of Arizona; 177: N/J-C/ESA – D Ducros; 178/179: ESA/N/J-C/Univ of Arizona; 180 & 181: N/J-C/Space Science Institute; 182/183: N/J-C; 184 left: N/J-C/ESA/SSI/Cassini Imaging Team; 184 right: N/J-C/SSI/Kevin M Gill; 185 left: N/J-C/Space Science Institute; 185 right: N/J-C/SSI/Gordan Ugarkovic; 186 & 188: N/J-C; 189: NASA/CXO/Univ College London/W Dunn et al; Optical: W M Keck Observatory; 190/191: Vzb83 on the basis of NASA pictures; 192/193: N/J-C/USGS; 194 & 195: N/J-C; 196: Voyager 2/NASA; 197: N/J-C/USGS; 198: N/J-C; 199: N/J-C/USGS; 200: Gavin James; 202: ESO; 203: ESA; 204: N/J-C/UMD; 205: ESA/Rosetta/NAVCAM (CC BY-SA IGO 3.0); 206: ESA/Rosetta/MPS for OSIRIS Team MPS/UPD/LAM/IAA/SSO/INTA/UPM/DASP/IDA; 207: N/J-C/USGS; 208 left: N/J-C/JHUAPL; 208 middle: NASA/Goddard/Univ of Arizona; 208 right: ESA/MPS for OSIRIS Team MPS/UPD/LAM/IAA/RSSD/INTA/UPM/DASP/IDA; 209: N/J-C/UCLA/MPS/DLR/IDA; 210: NASA/JHUAPL/SwRI; 211 (dotted zebra/Alamy); 212-220: NASA/JHUAPL/SwRI; 221: NASA/JHUAPL/SwRI/Roman Tkachenko